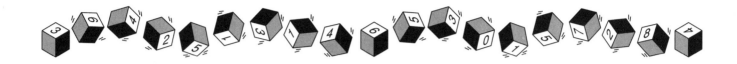

MEGA-FUN
MATH GAMES

70 Quick-and-Easy Games to Reinforce Math Skills

by
Dr. Michael Schiro

S C H O L A S T I C
PROFESSIONAL **B**OOKS

New York • Toronto • London • Auckland • Sydney

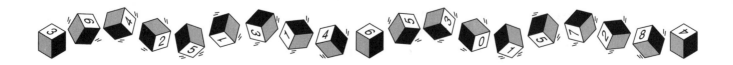

In memory of
Mary Schiro

Cover design by Vincent Ceci

Cover Illustrations by Anna Walker

Interior design and illustrations by Ellen Matlach Hassell
for Boultinghouse & Boultinghouse, Inc.

ISBN 0-590-48176-2

12 11 10 9 8 7 6 5 4 3   5 / 9

Printed in the U.S.A.

Contents

Introduction

Welcome to *Mega-Fun Math Games!* The games in this book are designed to help children learn and remember arithmetic facts, skills, and concepts by participating in enjoyable game-based activities. The games incorporate the ways children best learn mathematics: through the use of physical manipulatives within the context of developmentally appropriate practice.

The games are built around inexpensive and readily available materials: egg cartons, paper and pencils, tongue depressors, blank cards, wood cubes or dice, and paperboard. They are quick and easy to make, too.

The games are grouped by the materials used to make them and deal with a wide range of arithmetic topics, from mastering basic addition and multiplication facts to working with fractions and decimals. A Game Reference Chart on pages 9 and 10 makes it easy to locate games related to each arithmetic operation.

The games in this book will help students to develop their problem-solving skills, such as making and testing hypotheses, creating strategies, and organizing information. Plus as children play, they further their development of hand-eye coordination, concentration levels, visual discrimination, memory, and their ability to communicate and use mathematical language. On the social level, game playing can help children learn to work cooperatively, give and take praise and criticism, instruct others, and accept success and failure in the presence of peers. Watching children play mathematical games can also serve as a valuable assessment tool for teachers and parents.

Practical Considerations

Who Goes First

A fair way to pick the first player is with a "starting game." Common starting games involve flipping a coin, tossing dice, reciting counting-out rhymes, spinning a bottle, picking cards from a hat, or cutting a deck of cards. Besides choosing the first player (and if necessary, the second, third, and so on), starting games can be used to divide the class into teams and to designate any prized position in a game, such as team captain, dealer, or banker.

A Letter Home to Parents

Sending a short note to parents that tells them how and why mathematics games will be used in your classroom is a great idea. It can allay any doubts that may arise when their children come home describing how they "played games during math today!" Sending a game home with children to play with parents as part of homework is also useful. A number of books and articles describe how to host a Family Math Night where parents can play math games (as well as participate in other types of mathematic activities) with their children. This type of program helps parents get to know a school's mathematics program and gives them a sense of what can be learned from mathematics manipulatives and activities that are not textbook- or worksheet-based.

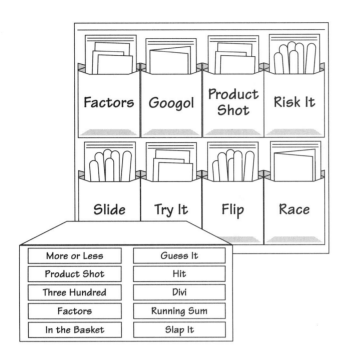

| Factors | Googol | Product Shot | Risk It |
| Slide | Try It | Flip | Race |

More or Less	Guess It
Product Shot	Hit
Three Hundred	Divi
Factors	Running Sum
In the Basket	Slap It

Storing Games

It is advisable to store math games in a single place, so that children know where to find and return them. Individual games can be stored in cardboard boxes (shoe boxes or pizza boxes work well). These can, in turn, be placed in larger cardboard boxes. Label each box with the name of the game (and decorate it, if desired) and glue the instructions, as well as a list of the materials used in the game, to the inside lid of the box. Games can also be stored in zippered plastic bags that are placed in hanging shoe bags or file folders that are stored in large cardboard boxes.

Rules for Using the Games

Encourage students to learn these three rules.

1. Before using a game, make sure all necessary materials are in the game's container. If not, report missing pieces to your teacher.

2. After using a game, put all of the materials back in the container.

3. After using a game, return it to its proper storage place.

Recording Sheets

Recording the problems solved while playing a math game can leave a mathematical trail that is of great value to children, teachers, and parents. Children can feel a sense of accomplishment as they look back at all of the math work they have done, teachers can use the records for assessment, and parents will appreciate this "evidence" that their children are actually doing mathematics and not just playing games. The illustrations below show several ways of structuring recording sheets. Note that children should not always be required to keep a written record. However, periodically requiring one is worthwhile.

Rectangles		
length	height	score

More or Less		
Round	Problem	Score
1		
2		
3		
4		
5		

Bounded Playing Areas

Use a piece of inexpensive felt as a playing surface to limit the area where a game's playing pieces are used. To confine the area where cubes (or other materials) are rolled, use a small box with felt glued to its bottom. Felt also serves to quiet the noise of playing pieces hitting the playing surface.

Introducing and Modeling Games

One of the most effective ways to teach children a new game is by modeling it for them. This can be done by playing for both opponents, playing against students while guiding their moves, or teaching one group of children how to play the game and then having them demonstrate it to others.

Ending a Game

If children are working in independent groups, they may need to be reminded when to finish playing their games. Telling children when the activity will end before they begin playing can help them finish on time. Or, try setting a sand timer and announcing that there are three minutes left until the end of an activity.

Groups

Groups of children can play the same game or different games. Depending on grouping arrangements, you'll need to construct different numbers of copies of each game. For example, some groups can play one game while others play another, and the groups can then switch games. In addition, small groups of children can play different games at selected times of the day while other activities take place. Use an activity rotation chart (such as the one below) to designate who plays which games at a certain time of day. Insert cards with children's names and game names into a pocket chart or affix to a Velcro board, or write names on a blackboard.

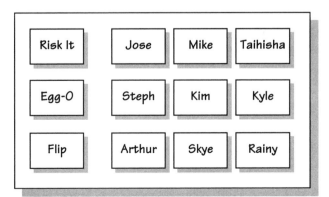

Competitive Versus Noncompetitive Games

Most of the games in this book have been designed as competitive games where the high scorer wins. All can be transformed into games where the high scorer is not the winner or into noncompetitive games. Most of the games in this collection can be played in such a way that players keep track of their own individual score over a period of days and try to better their previous day's scores. Children can enjoy keeping graphs of this information themselves.

Communication During Gaming

Many people think that a quiet room is one in which learning is taking place. However, when children are playing games in cooperative groups, they need to be able to talk with each other. This talk can be very constructive if children take the responsibility to make sure that all players in a game understand the algorithms, concepts, and facts being used within a game (see "Gaming Etiquette and Cooperative Learning").

Calculators

Calculators can be quite helpful for settling questions or disputes about answers, executing complex calculations, or keeping track of players' cumulative scores. Use your judgment as to whether calculators will speed up or defeat the purpose of a game.

Assessment

Adults who observe and interact with children while they are playing mathematical games can diagnose a wide variety of their mathematical strengths and weaknesses, notably arithmetic abilities, reasoning and problem-solving skills, and part-whole and figure-ground skills (seeing how the rules of a game work together, seeing how the parts of a problem relate to its solution, being able to focus attention on relevant parts of a game, etc.). In addition, the recording sheets that children produce while playing games can be placed in assessment portfolios, where they can be of great value to children, teachers, and parents (See, "Recording Sheets," on page 6.) Finally, games provide children with a powerful way of assessing their own mathematical abilities. The immediate feedback children receive from their peers while playing games can help them evaluate their mathematical concepts and algorithms and revise inefficient, inadequate, or erroneous ones.

Gaming Etiquette and Cooperative Learning

Much has been written about how to help students learn together in groups. The following guidelines for gaming etiquette will help students successfully play mathematical games. You might want to post them in your classroom and discuss them with your students.

➤ Play games to learn mathematics and have fun, not just to win.

➤ Teach each other and learn from each other. Every member of a group has the responsibility of helping the other members understand the rules of the game.

➤ Help your group to be successful. Work together to make the game an enjoyable learning experience for all group members. Do your fair share of the work.

➤ Never insult another player. Disagree with answers and ideas, not people. Be sensitive to the feelings of others.

➤ Take turns. Listen to what others have to say and think about it before responding.

➤ Ask an adult for help only when the group cannot help itself.

➤ Speak and act in ways that will not disturb others.

Varying Games to Meet Needs

Vary the games within this book to suit the needs of the children who will play them. Some ways of varying most of the games are described. The operations used within the games can be changed, the types of numbers used within the games can be changed, and the rules of the games can be changed. Please be creative in transforming the games into new forms, and please allow children to do likewise.

Finally, it is hoped that wherever and however the games are played, they will be truly enjoyed. They were devised as teaching tools and are practical and effective in that role. However, they need not always be used so purposefully. Parents and teachers can employ some of them as a substitute for a few hours of television, friends and siblings can add them to rainy-day repertoires, and several may occupy the more tedious hours of an automobile trip or school field trip.

Game Reference Chart

■ shows skill emphasis of main game

▼ shows skill emphasis of variations

	Page	Addition	Subtraction	Multiplication	Division	Place Value	Fractions	Inequalities	Multiples/Factors
1005 Count Up	12	■	▼			■	▼		
Capture the Fort	13		■					■	
Foreheaded	14					■			
Googol	15					■		■	
Number-Tac-Toe	16	■		■					
Ask and Give	18	■	■			■			
Hangmath	19	■	■	■	■	■			
Factors	20	■		■	■				■
Get One	22		■						■
Write It	23	■		■	■				■
Egg-O	25	■	■	■		▼	▼		
Egg Race	26	■		▼					
More or Less	27	▼		■				■	
Egg Bump	28	■	■						
Egg-a-Round	29	■		▼			▼		
Place-an-Egg Value	30	■				■		■	
Bing-Egg-O	32	▼		■			▼		
Three Hundred	33	■	■			■			
Egg Throw	35	■		■		▼			
Divide and Move	36				■				
Nines	39	■	■						
Roll It	40	■	▼	▼	▼				
Cardinal-Ordinal	41	■				■			
Product Shot	43	▼		■					
Rectangles	45			■					
Off the Board	47	■	■						
Number Grid	49	■				■			
Try It	50	■							
Numbers	52	■	■	■	■	■	■		
What's Up?	53				■				
Risk It	56	■		▼			▼		
Pick	57	■	▼	▼	▼				
Right On	58	■					▼		
Flip	59	▼		■				■	
Slide	60	▼		■					

Game Reference Chart

■ shows skill emphasis of main game

▼ shows skill emphasis of variations

	Page	Addition	Subtraction	Multiplication	Division	Place Value	Fractions	Inequalities	Multiples/Factors
Togol	61					■		■	
Guess It	62							▼	■
Get Seven	63	▼	▼	▼	■		▼		
Equato	64	■	■	■	■				
Ladders	65					▼	■	■	
Operations War	67	▼	▼	■				■	
Bongo	68	▼	▼	■	▼		▼		
Little Shot	69	■							
Place-a-Bet	71					■		■	
Running Sum	73	■							
Slap It	74	▼	▼	■	▼		▼		
Call It	75	▼	▼	■			▼		
What's Next?	76	■	■	■	■	■	■		
Two Numbers	77	■	■	■	■				
Mathino	78	■	▼	▼	▼				
The Great Trading Game	80	■	▼			■	▼		
0 to 99 Guess	82					■		■	■
In the Basket	84	■							
Race to the Nineties	86	■				■			
Checker Math	88	■	▼	▼	▼				
Down the Tubes	90	■	■			▼			
Hit	92	▼		■		▼	▼		
Drop the Die on the Donkey	94	▼		■					
Divi	96				■	▼			
Race	98	■	■	■	■		▼		▼
Jump the Answer	101	■	■	■	■				
Math Ball	102						▼		■
"Simon Says" Math	103	■	■	■	■	▼			
Number Calisthenics	104					▼			
Fingers	105	■	▼	▼					
Hands In	106	■							
Twist-'em	107	▼	▼			■			
Math Rover	108	■	■		▼		▼	▼	
Operation Hopscotch	110	■		▼					
Bean Bag Toss	111	▼		■	▼				

Paper-and-Pencil Games

Paper-and-pencil games are popular among children and require little preparation. The games in this book offer new ideas and variations on old ones. And they're not limited to pencil and paper! Let your students try them out on a chalkboard with chalk, on a sidewalk with chalk, or on a laminated piece of paper or cardboard with crayons. Doing so can help avoid the use of large amounts of paper.

1005 Count Up

SKILL AREAS:
counting, place value,
writing and saying numbers

Object: Players take turns writing and saying the numbers from 985 to 1005. Each player may claim one or two numbers in sequence, starting where the other player left off. The player who claims 1005 wins.

Number of Players: two

Materials: paper and pencil

Preparation: No preparation needed.

✎ Playing

1. The players take turns writing and saying the numbers from 985 to 1005 on a sheet of paper. The numbers are written in sequential order.

2. The first player starts by writing and saying either the number 985 or the numbers 985 and 986. Thereafter, each player may write either one or two numbers in sequence, starting from where the other player left off.

✎ Winning

The player who writes and says 1005 wins.

985
986
   987
988
   989
   990
991
992
   993
994
   995
996
   997
   998
999
1000
   1001
1002
1003
   1004
   1005

✎ Playing Variations

➤ Instead of using the numbers from 985 to 1005, the players can use the numbers from x to n, where x and n are any counting numbers. The game from 1 to 20 is excellent.

➤ Instead of writing either one or two numbers in sequence, the players can write up to k numbers in sequence, where k is any counting number between 2 and 10.

➤ The player who says the last number can lose rather than win.

✎ Skill Variation

Fractions: Instead of using counting numbers, players can use a sequence of fractions (for example, the fractions from ¼ to 5 with a counting increment of ¼ can be used, as in ¼, ½, ¾, 1 and so on).

Note: This game has a winning strategy that children can easily discover if they search for it.

Capture the Fort

SKILL AREAS:
subtraction, inequalities

Object: Each player has 50 points to bet during a series of rounds. Whoever bets the most points wins a round and advances.

Number of Players: two

Materials: paper, pencils, and a counter (such as a coin or button)

Preparation: Have the players draw a larger version of the playing board shown here on a sheet of paper.

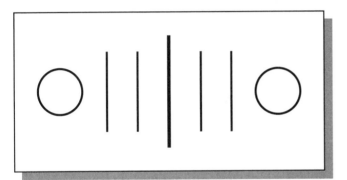

✎ Playing

1. Place the playing board between the two players. The circle nearest each player is that player's fort. The five lines between the circles are the positions on which the players can move the marker. The thick center line is the boundary between the players' territories.

2. The game starts with the marker on the center boundary line. Subsequently the marker is moved in either direction, one line at a time. The object is to move the marker into the opponent's fort, and thus capture it.

3. Moves are determined by making bets. Each player has 50 points to bet for the entire game.

4. The game is played as a series of rounds. During a round each player decides how many points to bet and secretly writes that number on a piece of paper. The players then show each other their bets. The player with the larger bet wins the round and moves his or her marker one space toward the opponent's fort. If both players bet the same number, the marker is not moved. A new round then begins.

5. Players use up betting points as the game proceeds. Before play begins, they each write "50" at the top of a small piece of paper. They subtract their first-round bet from 50, their second-round bet from the remainder of the first subtraction, and so on.

6. If a player runs out of points, he or she must bet zero points during each remaining round of the game, until the opponent either wins or runs out of points.

✎ Winning

The first player to move the marker into the opposing fort wins. If neither player has sufficient points left to reach the opponent's fort, the game ends in a tie.

✎ Playing Variations

➤ A player can be awarded a half-win if a tie results and the marker is in the opponent's territory.

➤ Players can avoid revealing their bets to each other by showing them instead to a third person, who announces only the winner of each round.

Foreheaded

SKILL AREAS:
reading, writing, and saying
numbers using place value language

Object: Each player writes a three-digit number on a slip of paper and tapes it to the forehead of the player on the left. Players then are given clues and try to guess their numbers.

Number of Players: three to six

Materials: pencils, paper, and tape (or sticky notes)

Preparation: Cut one 1-inch by 2-inch (3 cm by 5 cm) slip of paper, for each player. Small sticky back post-it tabs (about 1½ by 2 inches) are ideal.

✎ Playing

1. The players sit in a circle.

2. Each player secretly writes a three-digit number on a slip of paper and then tapes it to the forehead of the player to the left. (If sticky notes are being used, simply write on them and post them on each forehead.) The numbers are taped facing out so that every player sees all of the opponents' numbers, but not his or her own.

3. Players take turns in clockwise rotation.

4. During each turn, a player must guess the number on his or her forehead. If a player guesses correctly, he or she wins. If a player guesses incorrectly, opponents must tell him or her how many digits in the number are correct (the correct digit in the correct place). Then the player must give every opponent a clue. This involves telling each opponent a three-digit number, using place value terminology, that contains at least one digit that is the same and in the same position as a digit in the opponent's number. For example, a player might give the clue 135 to an opponent with the number 237 taped to her head. Players should write down the clues they are given, for future reference.

5. If one player gives another player a clue that was given on a previous turn by any player, that player must give another clue.

6. The game ends when a player guesses the number on his or her forehead.

✎ Winning

The winner is the first player to guess his or her number.

✎ Playing Variations

➤ Two-digit or four-digit numbers may be used instead of three-digit numbers.

➤ Foreheaded can be played so that a player is told both how many digits are completely correct (correct digit in the correct place) and how many digits are partially correct (correct digit in the wrong position). For example, if a player with 237 on her forehead guessed 532, she would be told, "One digit completely correct and one digit partially correct."

Googol

SKILL AREAS:
writing and reading numbers,
inequalities, ordering numbers

Object: One player writes eight numbers on slips of paper. Another player tries to guess which slip has the largest number. Points are awarded for guessing and ordering numbers.

Number of Players: two

Materials: paper and pencils

Preparation: No preparation needed.

✎ Playing

1. There are ten rounds in the game. During each round one player is the writer and the other is the picker. Players switch roles after each round.

2. To start a round, the writer tears a sheet of paper into eight small slips and secretly writes a different positive number on each one. The numbers may range from small fractions and decimals to a googol (a 1 followed by a hundred zeros). The writer places the slips face down on the playing surface.

3. During the round the picker turns the slips face up one at a time. The picker stops turning over slips when he or she believes that the largest number has been picked. The picker cannot go back to a previously turned-up slip; the decision to turn over another slip is final. If the picker turns over all eight slips, the last one turned over is considered to be the chosen one.

4. To end a round, the picker turns all eight slips face up and arranges them in order from smallest to largest.

5. Score each round as follows:
   ➤ three points to a picker who chooses the largest number
   ➤ one point to a picker who correctly arranges the slips in numerical order from smallest to largest

✎ Winning

The player with the highest total score after ten rounds wins.

✎ Playing Variation

➤ The following can be added to the scoring rules if writers incorrectly mark slips of paper: one point to the picker if the writer incorrectly marks the slips (a negative number, the same number twice, and so on).

Number-Tac-Toe

Object: Players add (or multiply) numbers to get sums (or products) equal to numbers on a tic-tac-toe grid. The first player to get three numbers in a row (as in tic-tac-toe) wins.

Number of Players: two

Materials: paper and pencil

Preparation: Create a Number-Tac-Toe grid by writing sums (or products) of the numbers from 1 to 9 in the nine cells of a regular tic-tac-toe grid (see sample, right) or use the grids on the following page.

X player				O player
1				1
2	5	7	12	2
3				3
4	16	9	15	4
5				5
6	2	6	8	6
7				7
8				8
9				9

✎ Playing

1. Players decide who will mark X's and who will mark O's. Each player lists the numbers from 1 to 9 in a column on either side of the grid. (See the sample playing sheet.)

2. The first player crosses out any one number in his or her column. Beginning with the second player, the game continues as follows.

3. During a turn a player crosses out any one number in his or her column of nine numbers that has not yet been crossed out. The player then adds (or multiplies) that number to (or by) the last number the opponent crossed out. If the sum (or product) is on the number-tac-toe grid and if it is not yet marked, the player marks an X or O over it.

4. The game ends when any of the following occurs:

   ➤ a player gets three marks in a row (as in tic-tac-toe)

   ➤ all of the numbers on the grid are marked "X" or "O"

   ➤ all nine numbers in each player's column of numbers are crossed out.

✎ Winning

The player who gets three marks in a row wins, as in tic-tac-toe. If neither player gets three marks in a row, a tie is declared.

Add-Tac-Toe Grids

5	7	12
16	9	15
2	6	8

14	3	12
8	16	6
2	18	10

7	9	15
16	11	5
3	13	4

6	4	10
12	14	8
16	9	5

13	8	9
15	5	11
17	10	7

15	12	9
16	7	5
2	8	6

Multiply-Tac-Toe Grids

25	15	56
12	36	16
49	20	8

24	6	56
8	35	14
63	8	15

8	10	5
48	24	3
36	6	30

12	24	4
35	27	18
15	21	5

3	36	63
15	30	9
18	24	8

15	12	36
16	25	49
20	8	56

Ask and Give

SKILL AREAS:
place value, reading and writing
large numbers, addition, subtraction

Object: In this variation of Go Fish, players use place value skills to trade numbers.

Number of Players: two

Materials: paper and pencils

Preparation: No preparation needed.

	Player A		Player B
A says, "Give me your 2's."	6 2 1,845		297,6 13
	+ 200,000	←	− 200,000
	8 2 1,845		97,6 13
B says, "Give me your 8's."	− 800	→	+ 800
	8 2 1,045		98,4 13
A says, "Give me your 4's."	+ 400	←	− 400
	8 2 1,445		98,0 13

✎ Playing

1. At the top of separate sheets of paper each player secretly writes a six-digit number, containing no zeros and no identical digits. Players keep their papers and numbers hidden from each other for the entire game.

2. As in Go Fish, players take turns being asker and giver. Each player tries to increase his or her number by taking digits from the other.

3. A turn begins when the asker says: "Give me your X's," where X can be any digit from 1 to 9 (for example, "Give me your 6's.").

4. If that digit is in the giver's number, the giver announces its place value (for example, "You get 600."). If that digit is not in the giver's number, the giver announces this (for example, "You get 0."). Note that the value of a digit that is asked for depends on its position in the giver's number. For example, if 6 is asked for, and the giver's number is 512,639, then the giver responds "You get 600." But if the giver's number is 561,243, then the giver responds "You get 60,000."

5. As soon as the giver responds with a number, the asker adds that amount to his number (for example, +600) and the giver subtracts that amount from his number (for example, −600).

6. Each player's number changes with each new addition or subtraction. Players always use the most recent form of their numbers when adding, subtracting, or announcing the positional value of a digit. Players keep track of their changing number by adding to and subtracting from their original number and its successors directly under the original number (see the sample game above).

7. If the same digit appears two or more times in the giver's number during play, the giver may announce either of its values. For example, in 621,063 the giver may say "You get 60," and say nothing about the 600,000.

8. The game ends after each player has had five turns as asker. Players then check each other's addition and subtraction.

✎ Winning

The player with the largest number at the end of the game wins. If either player's paper shows an error, that player automatically loses.

✎ Playing Variation

➤ This is an excellent game for calculator use.

Hangmath

Object: A variation of hangman. One player creates an arithmetic problem involving long addition, subtraction, multiplication, or division. The other player tries to reconstruct the problem before being hanged (within 14 guesses).

Number of Players: two

Materials: paper and pencils

Preparation: No preparation needed.

✎ Playing

1. Players take turns being hangman and guesser.

2. On one piece of paper the hangman secretly writes a long addition, subtraction, multiplication, or division problem.

3. On another piece of paper the hangman makes a playing board that shows the type of problem and position of the digits in the problem. Problems up to the following sizes are suitable: four-digit addition and subtraction; three-digit by three-digit multiplication; and two-digit into four-digit division.

4. The hangman hides the paper with the problem on it, but gives the guesser the playing board.

Hangman's Problem

$$
\begin{array}{r}
43 \\
\times\ 25 \\
\hline
215 \\
860 \\
\hline
1075
\end{array}
$$

Guesser's Playing Board

5. The guesser tries to reconstruct the problem by guessing which digits belong where on the playing board. Guesses have the following format: "Is there a ____ in the ____ column?"

6. If the guesser guesses a correct digit in the correct column, the hangman must indicate every place the digit occurs in that column.

7. If the digit guessed does not appear in the specified column, the hangman begins or adds a line to the picture of a hanged man. The hanged man consists of fourteen lines drawn in this order: base of gallows, upright post, cross beam, rope, head, neck, body, leg, leg, arm, arm, eye, eye, mouth.

8. As the game progresses, the guesser tries to use information obtained from previous guesses to guide further guesses.

9. The game ends when either the hangman completes the picture or the guesser reconstructs the problem. Then guesser and hangman switch roles.

Is there a five in the ones column?

✎ Winning

If the picture is completed, the hangman wins. If the arithmetic problem is completed, the guesser wins.

Factors

Object: Players take turns giving and receiving numbers to factor. They score points equivalent to the numbers given and the factors identified.

Number of Players: two (or whole class in two teams)

Materials: paper and pencil

Preparation: Have the players make a playing board, as shown below, or provide them with a copy of the playing board on page 21. The playing board has a scoring column for each player on opposite sides of the paper, and the numbers from 1 to 30 in an array in the center.

✎ Playing

1. Players alternate roles as picker and factorer.

2. On a turn as picker, a player crosses out any legal number on the playing board. (A legal number is any number that is not crossed out and which has *at least one* factor that also has not been crossed out.) The picker writes that number in his scoring column.

3. The factorer then crosses out any of the factors of the number the picker crossed out. The factorer writes all of these numbers in his scoring column. For example, if the picker crosses out 30 as the first play of the game, the factorer can cross out 1, 2, 3, 5, 6, 10 and 15. The factorer may choose to not cross out a factor of the number.

4. Neither player can reuse a number that has been crossed out.

5. Players switch roles after each round.

6. The game ends when there are no more legal numbers to cross out. Players then add the numbers in their columns to find their total score.

✎ Winning

The player with the highest total score wins.

✎ Playing Variations

➤ Have players keep cumulative sums of the numbers they acquire, rather than waiting until the end of the game to find the sum of the numbers.

➤ Use other numbers on the playing board; for example, 1 to 20, 1 to 35, 1 to 40.

➤ Allow the picker to cross out illegal numbers (numbers that have no remaining factors on the playing board). Add the rule that if the picker crosses out an illegal number, then the factorer (who now has no numbers to cross out) gets two consecutive turns as picker.

Emilio	Factors	Sonya
30	~~1~~ ~~2~~ ~~3~~ 4 ~~5~~	1
	~~6~~ 7 8 9 ~~10~~	2
	11 12 13 14 ~~15~~	3
	16 17 18 19 20	5
	21 22 23 24 25	6
	26 27 28 29 ~~30~~	10
		15

Factors

1	2	3	4	5
6	7	8	9	10
11	12	13	14	15
16	17	18	19	20
21	22	23	24	25
26	27	28	29	30

Factors

1	2	3	4	5
6	7	8	9	10
11	12	13	14	15
16	17	18	19	20
21	22	23	24	25
26	27	28	29	30

Get One

Object: Players subtract factors of a number from itself, until one player is left with a difference of 1.

Number of Players: two

Materials: paper and pencils

Preparation: No preparation needed.

✎ Playing

1. Players jointly choose a positive whole number to start the game.

2. The first player subtracts any factor of this starting number (except the number itself) from the starting number to get a difference. The second player then subtracts any factor of this difference (except the number itself) from the difference to get a new difference. If a prime number comes up during the game (having only factors of 1 and itself), the next player subtracts 1.

3. Players continue taking turns subtracting a factor of each new difference, always trying to leave the opponent a final difference of 1 (see the sample game).

Starting number	32
Player 1	− 8
	24
Player 2	− 12
	12
Player 1	− 6
	6
Player 2	− 3
	3
Player 1	− 1
	2
Player 2	− 1
Player 2 wins.	1

4. Both players do all subtracting on separate papers and check each other's work after each turn.

✎ Winning

The winner is the player who leaves the number 1 to the opponent.

Write It

Object: Each player secretly writes a number. If the sum of their numbers is a multiple of the predetermined number, then one player wins a point. If not, the other player wins a point.

Number of Players: two

Materials: paper and pencils

Preparation: No preparation needed.

✎ Playing

1. Before beginning, both players decide which whole number, X, the game will be about. (The best numbers to use are 2, 3, and 4.)

2. There are two sets of eight rounds in the game. During the first eight rounds, one player tries to get numbers to add up to a multiple of the chosen number X, and the other player tries to prevent this. The players switch roles for the second eight rounds.

3. To start a set of eight rounds, each player cuts a sheet of paper into eight smaller pieces.

4. During a round, each player takes a piece of paper and secretly writes a whole number on it. The numbers may be limited (for example, to those between 0 and 100). The players then show each other their numbers and add them together. If the sum is a multiple of X, the player trying for multiples of X scores one point; if not, the other player scores one point.

For example, John and Sue are playing a game about the number 2. John is trying to get numbers to sum to a multiple of 2 and Sue is trying to prevent this. John writes a 25 on his sheet of paper and Sue write the number 5 on her sheet of paper. The sum of the numbers is 30. John scores one point since the sum is a multiple of 2.

5. After the winner of a round is determined and the point awarded, a new round begins. After eight rounds, the players switch roles (the player who was trying to get numbers to sum to a multiple of X now tries to prevent this). Each player then cuts another sheet of paper into eight pieces and eight more rounds take place.

6. The game ends after two sets of eight rounds, when a total of 16 points have been awarded.

✎ Winning

The winner is the player with the most points.

✎ Skill Variation

➤ **Prime Numbers:** Limit the secret numbers to those between 1 and 40. The goal is to make the sum of the numbers be a prime number.

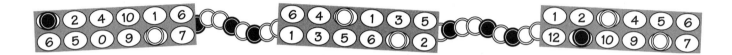

Egg-Carton Games

Give that empty egg carton a new life! Inside its lowly exterior lurks a randomizing device, a target, a racetrack, or a 3-D game.

Use the cartons that do not have large holes in their covers. Use either two-by-six cartons or three-by-four cartons. (The illustrations show two-by-six cartons; to use the three-by-four variety simply rearrange the numbers.) Do not remove the tops. Game instructions can be pasted inside the lids. Counters and other game pieces can be stored in the cartons.

When writing numbers in an egg carton, put them on both the bottom and the back side of the egg holes (so that the number in the hole can be read without removing objects from the egg hole). Buttons, centimeter cubes, pebbles, large beads, and other small objects are all appropriate to use as counters for egg-carton games.

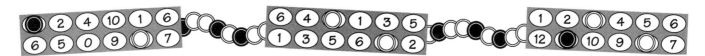

Egg-0

Object: Players take turns shaking a numbered egg carton containing two counters. The player's score after each turn is the sum, difference, or product of the two counters.

Number of Players: two to four

Materials: one egg carton and two small counters (buttons, pebbles, beads)

Preparation: Write the numbers from 0 to 10 in the holes of an egg carton, as shown. You will need to write one of the numbers twice.

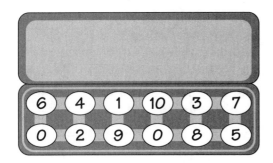

◯ Playing

1. Egg-O is played in rounds, with play rotating clockwise. During each round, each player takes a turn putting the two counters in the egg carton, closing it, shaking it, and then opening the carton to see where the counters landed.

2. In the addition game a player's score is the sum of the two numbers that the counters landed in. During a subtraction or multiplication game, players find either the difference or product of the numbers.

3. The player with the highest score wins one point for the round.

◯ Winning

The first player to acquire five points wins.

◯ Playing Variations

➤ Instead of acquiring points at the end of each round, players keep cumulative track of their scores from the rounds. The first player to get a cumulative score of 50 or more for the addition game wins (25 for the subtraction game, 200 for the multiplication game). Players can use calculators to help them keep track of their scores.

➤ Play ten rounds or for fifteen minutes. The winner is then the player with the highest cumulative score.

➤ Use three, four, or five counters, instead of two.

◯ Skill Variations

➤ **Fractions or Decimals:** Use fractions or decimals instead of whole numbers.

➤ **Place Value:** Use two-, three-, or four-digit numbers in the egg cavities, such as 25 or 378.

Egg Race

Object: Twenty numbers are generated from an egg carton to form a race track. Players race to see who can first correctly add another number to each of those in the race track.

Number of Players: two to whole class

Materials: one egg carton, one small counter (such as a button), paper, and pencils

Preparation: Write the numbers from 0 to 10 in the cavities of an egg carton, as shown. You will need to write one of the numbers twice.

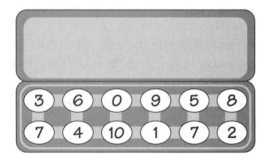

◯ Playing

1. To start, a player puts the counter in the egg carton, closes it, shakes it, opens it, and announces the number the marker landed on. This is done 20 times.

2. As the 20 numbers are announced, each player writes them down, in a straight column (using lined paper helps). The players then draw a line parallel to the column of numbers (see the sample game). These numbers now form a racetrack.

3. When all players are ready, a 21st number is generated by shaking the egg carton and is announced aloud. Players then race to see who can be first to add that number to each of the other 20 numbers on their racetrack. The players record each sum next to the

number on the other side of the line. They must work from the top number down, without skipping any numbers.

4. As players finish, they raise their pencils and yell, "Done!" The order in which the players finish is recorded. The player who finished first reads his or her answers. If they are all correct, that player wins. If not, the other players read their answers in the order in which they finished until a winner is found.

	+ 7 =
2	9
10	17
3	10
7	14
8	15
5	12
9	16
0	7
3	10
4	11
1	8
6	
5	
2	
7	

◯ Winning

The first player to correctly complete the addition race wins.

◯ Playing Variations

➤ Players can be provided with prenumbered playing boards.

➤ Students can play alone by keeping track of how long it takes them to complete a race. They can keep track of their scores and compare them over several days or weeks. Two students can take turns timing each other.

◯ Skill Variation

Multiplication: Instead of adding, players can multiply the 21st number by each of the other numbers.

More or Less

Object: Players bet on the size of the product of two numbers shaken in a numbered egg carton.

Number of Players: two or more

Materials: one egg carton, two counters, paper, pencils, and a playing piece for each player.

Preparation: Write the numbers from 0 to 10 in the egg carton. You'll need to write one number twice. Make a playing board or have players make their own by drawing two lines on a sheet of paper to make three columns labeled "Less than 20," "Equal to 20," and "More than 20."

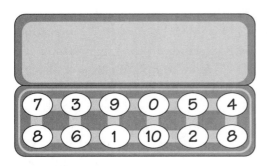

Less than 20	Equal to 20	More than 20

More or Less		
Round	Problem	Score
1		
2		
3		
4		
5		

◯ Playing

1. Have each player prepare a record sheet on lined paper that has columns labeled: Round, Problem, and Score. Number the lines in the Round column from 1 to 10.

2. Place a playing board within easy reach of each player. Choose a person to be the caller.

3. The game lasts 10 rounds, during which every player plays simultaneously.

4. At the beginning of each round, every player bets whether the product of the two numbers to be shaken by the caller will be more than, equal to, or less than 20. Players bet by placing their playing piece in the desired column of their playing board.

5. During each round, the caller places two counters in the egg carton, closes it, shakes it, opens it, and announces on which numbers the counters landed.

6. At the end of each round, players write the numbers announced by the caller in the Problem column of their record sheet in the form of a multiplication problem. They also secretly write down the answer to the problem. For example, if the caller announces "4 and 5," the players would write $4 \times 5 = 20$. Players then show each other their record sheets and calculate and record their scores. Players score one point for solving the arithmetic problem correctly and two points if they win their bet.

7. After ten rounds, each player adds his or her scores to get a total score.

◯ Winning

The player with the largest total score wins.

◯ Skill Variation

➤ **Addition:** Have children bet whether the sum of the numbers shaken is more than, less than, or equal to 10. Construct the playing board accordingly.

Egg Bump

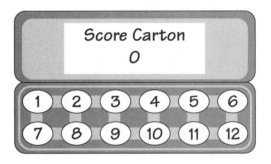

Object: Players shake two counters in a numbered egg carton, calculate the sum and difference of the resulting numbers, and record them in another numbered carton.

Number of Players: two

Materials: two egg cartons and 28 counters that can be divided into two sets (by color or shape) of 14 counters each

Preparation: Label the inside lid of one egg carton Shake Carton and write the numbers from 1 to 6 in its egg cavities, as shown. Label the inside lid of the other egg carton Score Carton, write the numbers from 1 to 12 in its egg cavities, and write the number 0 inside its top.

◎ Playing

1. Put the Score Carton between the two players. Give each player a set of 13 similar counters, and put the two remaining counters in the Shake Carton.

2. The game lasts 10 rounds. Each player takes one turn during each round.

3. To begin a turn, a player closes the Shake Carton, shakes it, and opens it to see where the two counters landed. The player calculates the sum and the difference (as a positive number) of the two numbers.

4. At the end of a turn, the player puts his or her counters in the Score Carton: one on the number corresponding to the sum that was calculated, and the other on the number corresponding to the difference that was calculated. If one of the player's own counters is already on one of these numbers in the Score Carton, the player leaves the original counter in place and does not place another counter on it. If one of the opponent's counters occupies the space, the player removes the opponent's counter, gives it back to the opponent, and replaces it with his or her own counter. Only one counter can ever cover a number in the Score Carton.

5. After 10 rounds (20 turns), the game ends and the players count the number of their counters in the Score Carton.

◎ Winning

The player with the most counters in the Score Carton wins.

◎ Playing Variation

➤ At the end of the game players find the sum of all of the numbers on which their counters reside. The player with the largest sum wins.

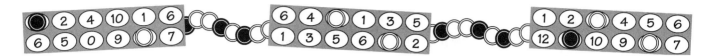

Egg-a-Round

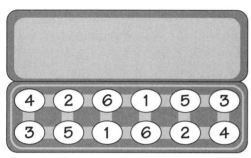

Leader's Carton

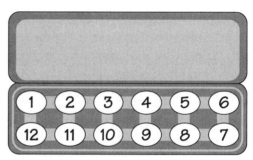

Player's Carton

Object: Players move their counters clockwise around a numbered egg carton according to the shake of a number-generating carton. They earn points according to the numbers on which their counters land.

Number of Players: two to whole class

Materials: one more egg carton than there are players and as many small counters (such as buttons) as there are egg cartons

Preparation: In one egg carton (Leader's Carton) write the numbers from 1 to 6, as shown. In the other egg cartons (Players' Cartons) write the numbers from 1 to 12, as shown.

◯ Playing

1. Choose a leader (either the teacher or a student). The leader takes the Leader's Carton and one counter.

2. Each player takes a Player's Carton and one counter. Each player puts his or her counter on any one of the 12 numbers.

3. The game is played as a series of six rounds.

4. To begin each round, the leader places the counter in his or her carton, closes it, shakes it, opens it to see where the counter lands, and announces that number.

5. After the leader announces the number, all players move their counters that many spaces, clockwise around their cartons, from where their counter last rested. The number on which they land is their score for that round.

6. Players add their scores cumulatively from round to round.

◯ Winning

The player with the highest cumulative score after six rounds wins.

◯ Playing Variations

➤ Use other numbers in the egg cartons. Try large numbers, such as 78 or 365, or fractions, such as ½ or ¾.

➤ The winner is the first player to get a score greater than 50.

◯ Skill Variation

➤ **Multiplication:** A player's score for a round is the leader's number times the number on which the player's counter lands.

29

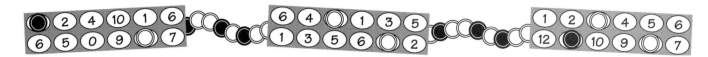

Place-an-Egg Value

> **SKILL AREAS:**
> place value, addition, inequalities

Object: Three counters designating hundreds, tens, and ones are shaken in a numbered egg carton. Players receive points for shaking large numbers and for reading and writing the numbers they shake.

Number of Players: two to four

Materials: one egg carton, and three counters you can write on (such as wood cubes or cardboard squares), pencils, and paper

Preparation: Write the numbers from 0 to 9 in an egg carton, as shown. You'll need to write two numbers twice. Mark the counters 100, 10, and 1. Provide players with recording sheets (on page 31).

◎ Playing

1. The game is played in five rounds. During each round, each player takes one turn, in clockwise rotation.

2. During a turn, the player places the three counters in the egg carton, closes it, shakes it, and opens it to see where they land. The player announces the number of hundreds, tens, and ones acquired (for example, 4 hundreds, 3 tens, and 5 ones). The player then says aloud and writes down the number in both expanded notation (400 + 30 + 5), and place value notation (435) on the recording sheet provided.

3. After each round, players compare their numbers. The player with the largest number receives one point. In addition, players who correctly read and write their numbers in expanded and place value notation get one point.

4. After five rounds of play, each player adds his or her points from all five rounds.

◎ Winning

The player with the most points wins.

◎ Playing Variations

➤ Write both one- and two-digit numbers in the egg carton so that players must regroup numbers when moving from expanded to place value notation.

➤ Leave counters blank and have players create the largest number possible by deciding, after shaking the carton, which counter should be 100, which 10, and which 1.

➤ Add another counter labeled 1,000.

◎ Skill Variation

➤ **Addition:** Have players add together the numbers they receive starting with the second round. The player with the largest sum at the end of each round gets 1 point.

Place-an-Egg Value

Round	Hundreds	Tens	Ones	Number	Score
1					
2					
3					
4					
5					

Final Score =

Place-an-Egg Value

Round	Hundreds	Tens	Ones	Number	Score
1					
2					
3					
4					
5					

Final Score =

Place-an-Egg Value

Round	Hundreds	Tens	Ones	Number	Score
1					
2					
3					
4					
5					

Final Score =

Bing-Egg-O

Object: Similar to Bingo. Egg cartons containing multiplication products serve as Bingo cards. Instead of picking Bingo numbers, the caller shakes two numbered egg cartons containing counters to obtain factors.

Number of Players: three to whole class

Materials: two more egg cartons than the number of players, and 11 counters for each player

Preparation: In two of the egg cartons write the numbers from 1 to 9, as shown below. In the other egg cartons write a random selection of the following numbers: 1, 2, 3, 4, 5, 6, 7, 8, 9, 10, 12, 14, 15, 16, 18, 20, 21, 24, 25, 27, 28, 30, 32, 35, 36, 40, 42, 45, 48, 49, 54, 56, 63, 64, 72, 81. These are the products of the numbers from 1 to 9.

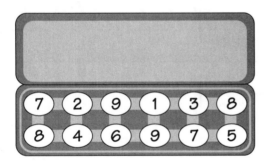

◯ Playing

1. Select a caller. Give the caller the two egg cartons numbered from 1 to 9 and two counters. Give each player one of the other egg cartons and 10 counters. Players open their cartons and place their counters in the lids.

2. The caller places a counter in each egg carton, closes and shakes them, opens them to see which numbers the counters landed on, announces the two numbers out loud, and writes them down.

3. The players silently multiply the two numbers together and look to see if the product

is written in their cartons. If it is, they place a counter on the number.

4. This continues until a player has counters either in a straight line of six or in two adjacent rows of three. The player then calls out "Bing-Egg-O."

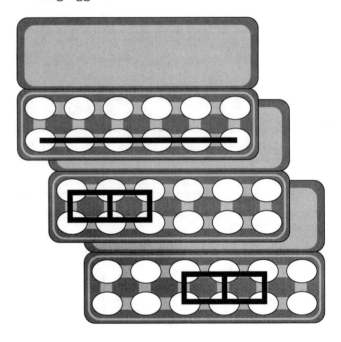

5. The caller checks to see that this player has covered products that correctly match the pairs of numbers shaken.

◯ Winning

The first player to validly call "Bing-Egg-O" wins.

◯ Skill Variations

➤ **Addition:** Write any whole numbers in the caller's egg cartons, figure out the sums, and write them in the players' egg cartons.

➤ **Fractions:** Same as above, but use fractions.

Three Hundred

Object: Players generate eight two-digit numbers, which they cumulatively add to or subtract from each other in an attempt to get a result that is as close as possible to 300.

Number of Players: two to whole class

Materials: one egg carton, two small counters, lined paper, and a pencil for each player

Preparation: Write the digits from 1 to 9 in the egg carton, as shown. You'll need to write three of the numbers twice.

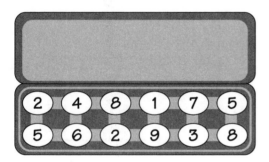

◯ Playing

1. Choose one person to be the shaker for the entire game. Distribute lined paper and a pencil to each player.

2. The game is played for up to 11 rounds. Players create a number as close as possible to 300 by the eleventh round.

3. To begin each round, the shaker puts the two counters in the egg carton, closes it, shakes it, opens it to see on which two digits the counters landed, and announces the digits. Each player writes down the two digits, in either order, to make a two-digit number. (For example, if counters land on 2 and 5, the players may create either 25 or 52.) In the first round, each player records the first two-digit number at the top of the paper.

4. During the second round, the players record their second two-digit number directly under

the first number and either add or subtract. The result is written under the two numbers.

5. In succeeding rounds, players record the two digit number directly below the cumulative result of the previous round, either add or subtract, and record the result below the problem.

6. During the game players may add no more than eight times and subtract no more than two times. (Eleven rounds allow for the first number, eight additions, and two subtractions.) Each player decides when to add or subtract during the game.

7. At any time during the game, before the beginning of a new round, players can declare that they are satisfied with their cumulative result of the previous round and that they wish to stay with that result for the remainder of the game. Players do this if they feel that their cumulative result is as close to 300 as possible.

	2 5
+	6 3
	8 8
+	4 2
	1 3 0
+	5 6
	1 8 6
+	4 9
	2 3 5
+	2 7
	2 6 2
+	7 7
	3 3 9
−	8 9
	2 5 0
+	8 2
	3 3 2
−	5 5
	2 7 7
+	2 5
	3 0 2

◯ Winning

The player who creates the number closest to 300 (either above or below) wins.

◯ Playing Variation

➤ Provide players with copies of the recording sheets on the next page. They can be helpful during the first few times the game is played.

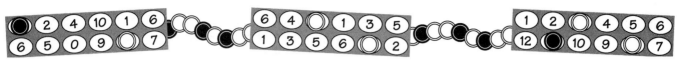

Three Hundred

+ + + + + + + + − −

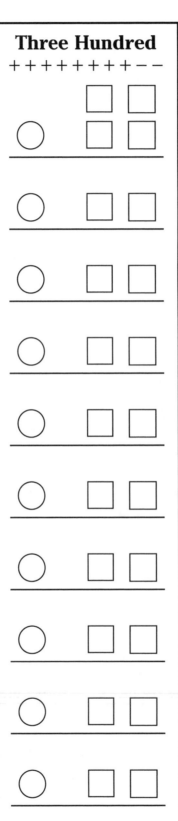

Three Hundred

+ + + + + + + + − −

Three Hundred

+ + + + + + + + − −

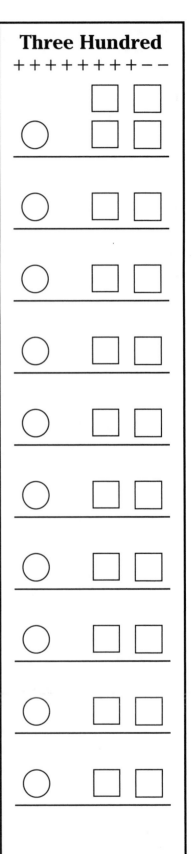

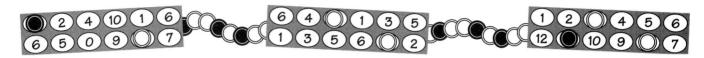

Egg Throw

Object: Players take turns throwing numbered cardboard squares at an egg carton target. They score points by multiplying the numbers on the squares by the numbers on which they landed in the egg carton.

Number of Players: two to four

Materials: one egg carton, and one piece of cardboard

Preparation: Cut four cardboard squares about ¾ inch (2 cm) on a side. Number them to correspond to the multiplication facts to be drilled. For example, to drill the four, five, six, and seven facts, number the squares 4, 5, 6, and 7. Write the numbers from 0 to 10 in the egg carton, as shown.

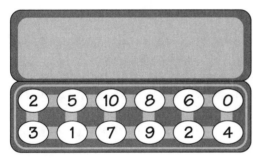

◯ Playing

1. Place the egg carton on the floor. Draw a tossing line behind which players must stand when tossing.

2. The game is played as a series of five rounds. Every player takes a turn during each round.

3. To begin a turn, a player stands behind the tossing line and tosses the four squares at the egg carton, one at a time.

4. To end a turn, a player calculates his or her score by multiplying the number on each square by the number on which it fell. (Squares which do not fall into egg cavities do not score points.) Then he or she adds the four products to get a score for the round. Finally, the new score is added to scores from previous rounds. (Calculators can be used to check answers or to keep track of cumulative scores.)

◯ Winning

The player with the largest total score after five rounds wins.

◯ Playing Variation

➤ Provide players with recording sheets.

Round	Calculations	Score	Cumulative Score
1			
2			
3			
4			
5			

◯ Skill Variations

➤ **Place Value:** Write the numbers 1000, 100, 10, and 1 on the squares. At the end of each turn players record and read their **scores** using place value language.

➤ **Money:** Use a penny, nickel, dime, and quarter instead of cardboard squares. The number in the egg cavity in which a coin lands indicates how many of that coin the player receives. Players calculate their scores using money language and notation.

➤ **Addition:** The number on each square is added to the number on which it falls.

Divide and Move

Object: Players move along a racetrack containing the numbers from 10 to 90. They move according to the remainders that result from dividing their playing position number by a number generated from an egg carton.

Number of Players: two to six

Materials: one egg carton, one counter, lined paper, and pencils

Preparation: Write the numbers from 1 to 10 in the egg carton. You'll need to write two numbers twice. Have each player write the numbers from 10 to 90 on lined paper, leaving about two inches between columns, or use photocopies of the playing board on the next page.

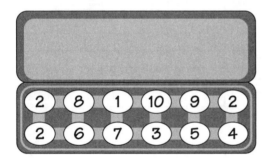

◯ Playing

1. The players circle the number 10 on their paper.

2. Players take turns with play rotating clockwise. A turn involves three steps.

3. First, a player puts the counter in the egg carton, closes it, shakes it, and opens it to see which number the counter landed on.

4. Next, the player divides that number into the largest number circled on his or her paper and finds the remainder for this division problem.

5. The remainder determines how many more numbers will be circled on the player's paper. The player circles numbers successively, beginning with the one after the previously circled number. For example, if on the first turn the remainder is 2, then the player circles 11 and 12 (the first two numbers after 10). Note that a remainder of 0 means that no numbers get circled that turn.

6. A player who notices that another has made a mistake gets to take that player's turn.

◯ Winning

The first player to circle all the numbers up to 90 wins.

◯ Playing Variation

➤ To play the game more quickly, use fewer numbers.

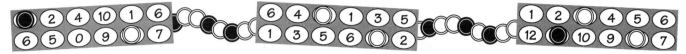

Divide and Move

10			
11	31	51	71
12	32	52	72
13	33	53	73
14	34	54	74
15	35	55	75
16	36	56	76
17	37	57	77
18	38	58	78
19	39	59	79
20	40	60	80
21	41	61	81
22	42	62	82
23	43	63	83
24	44	64	84
25	45	65	85
26	46	66	86
27	47	67	87
28	48	68	88
29	49	69	89
30	50	70	90

Cube Games

One common way to generate random numbers from 2 to 12 is by rolling two dice. The games in this section, however, call for numbers anywhere from –25 to 100! You can create your own number cubes by covering the faces of regular dice with masking tape and writing on the new symbols, or by marking blank cubes purchased from an educational materials supply company, or by using the cube pattern provided on page 112.

Other easily-made devices that also randomize six numbers will work here as well. For more ideas see page 112.

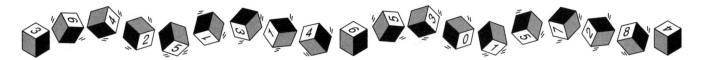

Nines

SKILL AREAS:
addition, subtraction, probability

Object: Players cross out the digits from 1 to 9 according to the sum rolled on two numbered cubes. The player who crosses out the most numbers wins.

Number of Players: one to four

Materials: two cubes, paper, and pencil

Preparation: On the faces of each cube write the numbers from 1 to 6.

🎲 Playing

1. Nines is played in turns, with each player completing a turn and scoring before the next player's turn begins.

2. A player begins a turn by preparing a playing board. This involves writing the numbers from 1 to 9 on a piece of paper, as in the sample game.

3. During a turn a player rolls both cubes, adds the numbers rolled, then crosses out one or more numbers on the playing board equal to the sum of the numbers rolled. For example, if the players rolls a 4 and a 5, their sum is 9, and the player can cross out any of the following groups of numbers: 9; 8 + 1; 7 + 2; 6 + 3; 5 + 4; 6 + 2 + 1; 5 + 3 + 1; or 4 + 3 + 2. (See the sample game.) The player continues the process of rolling cubes and crossing out numbers until a sum is rolled that cannot be crossed out on the playing board.

4. A player's score may be either the number of digits not crossed out or the sum of the digits not crossed out. (Decide which before the game begins.) After a player's score is calculated, the next player's turn begins.

🎲 Winning

The player with the lowest score wins.

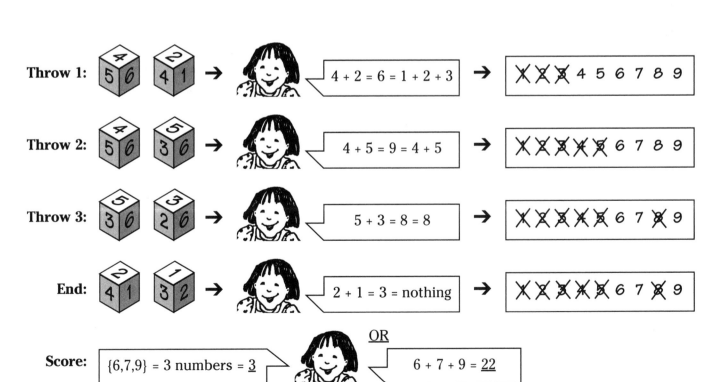

Throw 1: $4 + 2 = 6 = 1 + 2 + 3$ → X̶ X̶ X̶ 4 5 6 7 8 9

Throw 2: $4 + 5 = 9 = 4 + 5$ → X̶ X̶ X̶ X̶ X̶ 6 7 8 9

Throw 3: $5 + 3 = 8 = 8$ → X̶ X̶ X̶ X̶ X̶ 6 7 X̶ 9

End: $2 + 1 = 3 = $ nothing → X̶ X̶ X̶ X̶ X̶ 6 7 X̶ 9

Score: {6,7,9} = 3 numbers = <u>3</u>   OR $6 + 7 + 9 = $ <u>22</u>

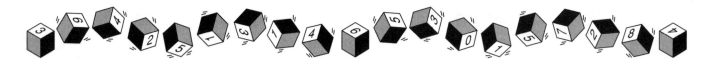

Roll It

Object: Five numbered cubes are repeatedly rolled, setting aside one cube after each roll, in an attempt to cumulatively roll the largest sum possible.

Number of Players: two to six

Materials: five cubes

Preparation: On the faces of each of the five cubes write the numbers 3, 4, 5, 6, 7, and 8.

Playing

Each player has only one turn during Roll It. On a turn a player follows these five steps:

1. First, the player rolls all five cubes and sets aside the cube showing the largest number.

2. Second, the player rerolls the remaining four cubes and again sets aside the cube with the largest number on it. The numbers on the two cubes set aside are then added and the sum announced.

3. The remaining three cubes are rerolled. The cube with the largest number on it is set aside, and its number is added to the previous sum obtained.

4. The remaining two cubes are rerolled. The cube with the largest number on it is set aside, and its number is added to the previous sum obtained from the first three cubes set aside.

5. Finally, the player rerolls the remaining cube, adds its number to the previous sum, and announces the result. This is the player's final score.

6. After one player's turn is completed, the next player's turn begins. When every player has taken one turn, the game ends.

Winning

The player with the largest final score wins.

Playing Variations

➤ Use fewer or more than five cubes.

➤ Write different numbers on the cubes, for example, the numbers from 1 to 6 or those from 24 to 29.

Skill Variations

➤ **Multiplication:** the numbers acquired in the game can be successively multiplied together to obtain a single product, which is a player's final score.

➤ **Multiple Arithmetic Operations:** Play the game by first adding, then subtracting, then multiplying, and then dividing the numbers thrown and set aside, in that sequence. (Decimals can be used or players can round off to the nearest whole number after dividing.) Or just three cubes can be used, with an addition and then a subtraction.

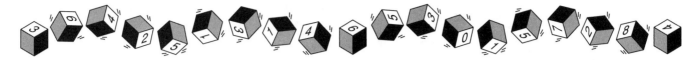

Cardinal-Ordinal

SKILL AREAS:
cardinal and ordinal
numbers, addition

Object: Players cross out 12 cardinal numbers and 12 ordinal numbers according to the numbers thrown on three cubes. The first player to cross out all 24 numbers wins.

Number of Players: two to five

Materials: three cubes, paper, and pencils

Preparation: Write the numbers from 1 to 6 on the faces of each cube. Have each player make a copy of the game sheet on the next page, or reproduce copies for each player.

◈ Playing

1. Each player attempts to be the first to cross out all of the number names in both the cardinal and ordinal columns of their game sheets. Players may cross out cardinal numbers in any order. However, they must cross out ordinal numbers in sequence from first to twelfth.

2. Players take turns in clockwise rotation. Players write on their own game sheet.

3. To begin a turn, a player rolls the three cubes simultaneously.

4. Next, the player crosses out number names according to the numbers rolled on the cubes. Number names may be crossed out if they match a number rolled on a single cube, the sum of the numbers rolled on two cubes, or the sum of the numbers rolled on all three cubes. A player may do any combination of these things on a single turn, so long as each number rolled is used no more than once. Number names in either column or both columns may be crossed out on a single turn.

For example, if on his or her first roll a player rolls a 1, 4, and 5, the player might cross out any one of the following:

- *one, four,* and *five*
- *first, four,* and *five*
- *ten* (1 + 4 + 5)
- *first* and *nine* (4 + 5)
- *one* and *nine*
- *four* and *six* (1 + 5).

A player does not have to use all of the numbers rolled: in the above example the player might have just crossed out either *one* or *one* and *five*. If a player is not able to use any cubes, he or she passes and the next player begins a turn.

5. A player's turn ends when he or she finishes crossing out number names and passes the cubes to the next player.

◈ Winning

The first player to cross out all of the cardinal and ordinal number names wins.

◈ Playing Variations

➤ Allow players to steal a number another player rolls, if the player who rolled the number can use it but does not do so. The stealer must claim the number as soon as the cubes are passed to the next player and must be able to use the stolen number immediately.

➤ Use two columns of cardinal number names, ordinal number names, or numerals from 1 to 12.

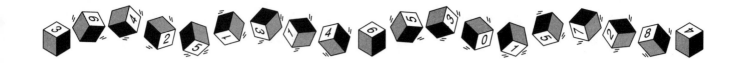

Cardinal (any order)	Ordinal (in sequence)	Cardinal (any order)	Ordinal (in sequence)
one	first	one	first
two	second	two	second
three	third	three	third
four	fourth	four	fourth
five	fifth	five	fifth
six	sixth	six	sixth
seven	seventh	seven	seventh
eight	eighth	eight	eighth
nine	ninth	nine	ninth
ten	tenth	ten	tenth
eleven	eleventh	eleven	eleventh
twelve	twelfth	twelve	twelfth

Product Shot

Object: A numbered cube is rolled 13 times. Players write the resulting numbers in the hexagons of a game sheet. Scores are calculated by multiplying together the numbers in adjacent hexagons.

Number of Players: two to whole class

Materials: one cube, and multiple copies of the game sheet

Preparation: Write the numbers 2, 3, 4, 5, 6 and 7 on the faces of a cube. Make a copy of the game sheet (on the next page) for each player.

◼ Playing

1. Give each player a game sheet and pencil.

2. Choose one person to roll the cube and announce the numbers rolled. This is done 13 times during the game.

3. Immediately after each number is announced, every player writes that number in any empty hexagon on their game sheet. Once written, a number's position on the game sheet cannot be changed.

4. When all 13 hexagons on the players' game sheets are filled, each player finds the products of the numbers in diagonally adjacent pairs of hexagons (indicated by the arrows on the game sheet). Each product is recorded in the appropriate circle.

5. Players cross out any product that appears in only one circle.

6. Players find the sum of the products that are not crossed out. These are their scores.

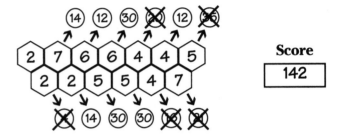

Score
142

◼ Winning

The player with the highest score wins.

◼ Playing Variations

➤ Have a player's score be simply the number of circles not crossed out.

➤ Drill other multiplication facts by writing different numbers on the cubes.

➤ Play three games. The winner is the player with the highest cumulative score.

◼ Skill Variation

➤ **Addition:** Add the numbers and cross out sums that appear in only one circle. A player's score is the number of remaining circles.

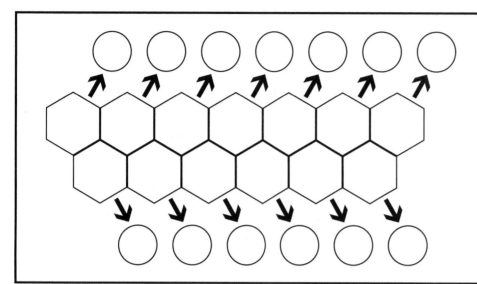

Product Shot

Score

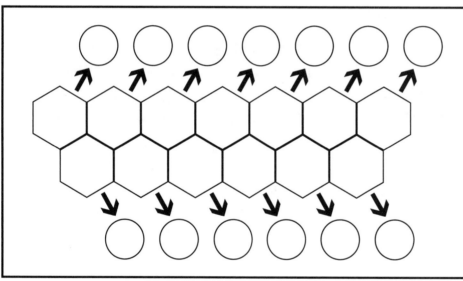

Product Shot

Score

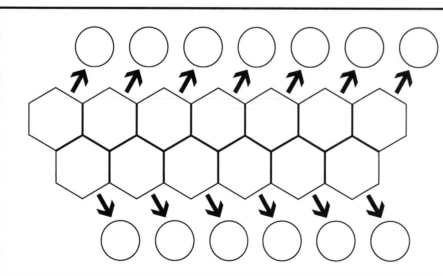

Product Shot

Score

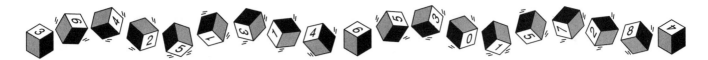

Rectangles

Object: During a series of rounds, players toss two numbered cubes that determine the length and width rectangles that are constructed on 12 by 12 pieces of graph paper. Points are scored by finding the areas of rectangles.

Number of Players: two to five

Materials: two dice or cubes, one pencil or crayon for each player, and one copy of the playing board on the next page (or graph paper) for each player

Preparation: Number each cube from 1 to 6 and duplicate copies of the playing board (or have players mark off a 12 by 12 square on a piece of graph paper). To make playing boards reusable, laminate the boards.

🎲 Playing

1. Players take turns. During a turn, a player tosses the cubes and constructs a rectangle by marking its length on the playing board (or graph paper) on a horizontal line according to the number thrown on one cube, and marking its height according to the number thrown on the other cube. The player then outlines the entire rectangle, colors it in, and calculates his or her score by determining the number of squares within the rectangle.

2. The rules for placing rectangles are as follows: All rectangles must be placed entirely within the major 12 by 12 square playing area; the edges of rectangles may touch (but do not have to); rectangles may not overlap each other; and no rectangle may be placed within another rectangle.

3. Players drop out of the game and calculate their cumulative score when their throw of the dice gives them a rectangle that will not

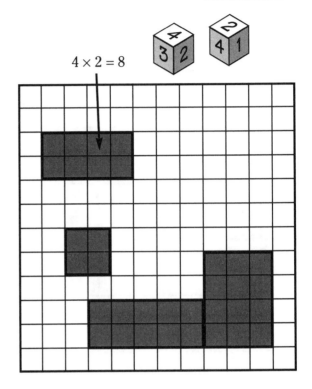

$4 \times 2 = 8$

fit on their playing board. The last player to place a rectangle on his or her graph paper gets a bonus of 10 points to add to their cumulative score. The game ends when all players have dropped out.

🎲 Winning

The player with the highest cumulative score wins.

🎲 Playing Variation

➤ Have players write down the method of calculation that they used to find the area of their rectangle under the rectangle. For example, for a 3 by 4 rectangle they might write $3 \times 4 = 12$, $3 + 3 + 3 + 3 = 12$, or $4 + 4 + 4 = 12$.

45

Rectangles

Off the Board

Object: Players move their counters in the positive or negative direction on a number line according to the throw of two numbered cubes. The first player to get a counter past either +7 or −7 wins.

Number of Players: two

Materials: two cubes, two different counters, and copies of the playing board

Preparation: With a blue pen write the numbers 0, 1, 1, 2, 2, 3 on one cube. With a red pen write the same numbers on the other cube. Make two larger-size copies of one of the playing boards on the next page. The first playing board is for beginners, and the second is for advanced players.

🎲 Playing

1. Give one copy of the appropriate level board to each player.

2. Each player chooses a counter and places it on the "Middle" (or 0) position on their playing board. Players move only their own counter.

3. Players decide which cube, the red one or the blue one, they will use to move their counters in the "up" (or positive) direction and which cube will move their counters in the "down" (or negative) direction.

4. Players take turns. During a turn a player rolls the two cubes and then moves his or her counter on the playing board, starting from where it last rested. The player moves the counter in the "up" (or positive) direction a number of spaces equal to the number rolled on the "up" (or positive) cube and then moves the counter the number of spaces in the "down" (or negative) direction indicated on the "down" (or negative) cube.

5. The players continue taking turns until one player's counter is moved off either end of the playing board, past either "up 7" (+7) or "down 7" (−7). The game then ends.

🎲 Winning

The first player to move his or her counter off either end of the playing board wins.

🎲 Playing Variations

➤ For younger players, color one side of the playing board red and the other side blue to correspond to the colors on the cubes.

➤ Require older players to correctly say or write the equations that correspond to each move.

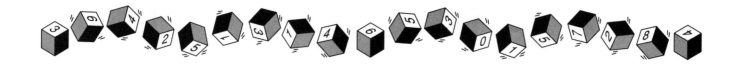

	win	
up		**7**
up		**6**
up		**5**
up		**4**
up		**3**
up		**2**
up		**1**
middle		
down		**1**
down		**2**
down		**3**
down		**4**
down		**5**
down		**6**
down		**7**

win

win
+7
+6
+5
+4
+3
+2
+1
0
−1
−2
−3
−4
−5
−6
−7

win

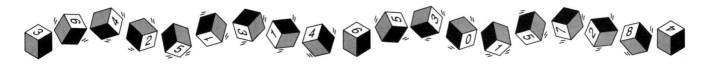

Number Grid

Object: Players create and write numbers in the cells of a 10 by 10 square grid, in an interlocking manner similar to Scrabble™.

Number of Players: two to five

Materials: three cubes, paper, and pencils

Preparation: On one cube write the numbers 1, 2, 3, 5, 7, and 9. On the second write the numbers 1, 2, 4, 6, 8, and 9. On the third cube write the numbers 3, 4, 5, 6, 7, and 8. Have the players draw a 10 by 10 square grid. (If possible, have them outline the grid on a piece of graph paper or draw vertical lines on lined paper.) The grid on page 46 may be reproduced.

◆ Playing

1. Players take turns in clockwise rotation. All players write their numbers on the same 10 by 10 grid. Each player records points scored on a separate sheet of paper.

2. To begin a turn, a player rolls the three cubes. The player then tries to use the three digits rolled to form a three-digit number that will fit on the grid and read either from left to right or top to bottom.

3. The player writes the number on the grid, one digit per square. The first number created is written anywhere near the center of the grid. Thereafter, **all numbers created must interlock with one or more previous numbers**, as in Scrabble™ or crossword puzzles. Either one or two digits of the number created may overlap, but not all three. Each play must add at least one new digit to the grid. Players need not worry about digits adjacent to the three-digit number they place on the grid, only that their number has either one or two digits in common with the configuration of numbers already on the grid. In the

example, the first number is 721. The second number, 374, interlocks with it in one place. The third number, 318, interlocks with the 3 in 374. Note that it is permissible for 318 to be adjacent to 721.

The fourth number is 819 and the fifth is 439; both interlock with the previous configuration in two different places.

The numbers 972 and 873 illustrate other ways that numbers can interlock. If a player cannot place a number on the grid, that player passes, and it becomes the next player's turn.

4. At the end of a turn, a player scores points equal to the number just placed on the grid. (For example, the player who created 374 gets 374 points.) Any points acquired are immediately added to points acquired from previous turns. Each successive score should be recorded in a straight column below the previous total. A player who cannot place a number on the grid scores 0.

5. The game ends either when the grid is filled or when each player has had to pass three times in succession.

◆ Winning

The player with the most points at the end of the game wins.

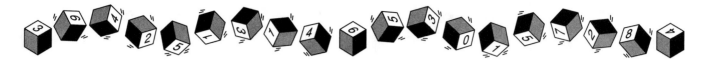

Try It

Object: Players bet whether or not a 2 will be rolled as they keep a cumulative sum of the numbers rolled on two cubes.

Number of Players: three to whole class

Materials: two cubes, paper, and pencils

Preparation: On each cube write the numbers from 1 to 6. Have each player prepare a playing sheet on lined paper or make copies of the reproducible record sheet for players.

Playing

1. Choose one person to be the roller for the entire game.

2. Try It is played in rounds. A round ends when a 2 is rolled on the cubes. Players try to get a total score of 100 or more points before the game ends.

3. During a round the roller rolls both cubes, announces the numbers rolled, the sum of the numbers rolled, and the cumulative sum of all the numbers rolled during the round. The cube roller repeats this until a 2 is rolled on a cube to end the round.

4. During a round the players do two things: They keep track of the cumulative sum of all numbers rolled during the round and they decide when to drop out. A player may drop out of a round at any time, and when he or she does, that player's round score is equal to the cumulative sum of all the numbers rolled during the round up to that point. Players announce when they drop out by recording their round score on their playing sheet, raising their pencil in the air, and saying, "I'm out!" Players must drop out between rolls of the cubes, not during a roll.

5. A round ends when the roller rolls a 2. A player who has not dropped out before a single 2 is rolled receives a round score of zero. A player who has not dropped out when a double 2 is rolled loses all points thus far accumulated during the game and receives a zero for both the round score and the total score.

6. When a round ends the players calculate their total score for the game thus far. The total score is the round score from the just-completed round plus the total score from the previous round (unless a player lost all points on the roll of a double 2). The players announce their "total scores" after calculating them. (In a large group, only the players with the highest scores announce them.) If one or more players has a total score of 100 or more, the game ends. If not, a new round begins.

7. The roller should pause between each roll of the cubes so that players who wish to drop out of the round have a chance to do so. The roller should also call on players to help figure out cumulative sums. At any time the roller may also ask all players who have not yet dropped out to raise a hand.

Winning

All players with a total score of 100 or more at the end of the game win. There can be more than one winner.

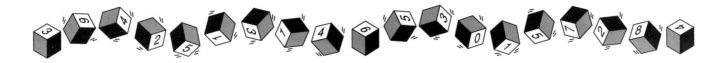

Try It

Round	Round Score	Total Score
1		
2		
3		
4		
5		
6		
7		
8		
9		
10		

Try It

Round	Round Score	Total Score
1		
2		
3		
4		
5		
6		
7		
8		
9		
10		

Try It

Round	Round Score	Total Score
1		
2		
3		
4		
5		
6		
7		
8		
9		
10		

Try It

Round	Round Score	Total Score
1		
2		
3		
4		
5		
6		
7		
8		
9		
10		

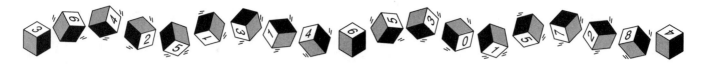

Numbers

Object: Five cubes with numbers and operations written on their faces are tossed. Players create as many different numbers as they can in three minutes using the numbers and operations tossed.

Number of Players: two to whole class

Materials: five cubes, paper, pencils, and a clock or egg timer

Preparation: Write the following numbers and operations on the cubes:

cube one: 4, 5, 6, −, ×, ÷

cube two: 0, 1, 2, 3, +, −

cube three: 0, 1, 2, 3, ×, ÷

cube four: 7, 8, 9, +, −, ×,

cube five: 4, 5, 6, +, −, ×

🎲 Playing

1. Choose a person to act as the roller. That person rolls all five cubes, announces the numbers and operations, and starts the egg timer or checks the clock.

2. After the dice are rolled, players write down as many numbers as they can create using the numbers and operations rolled.

3. A player does not have to use all of the numbers and operations rolled to create a number: one, two, three, four, or all five of the cubes rolled can be used. Players may combine the numbers and operations rolled in any way, if doing so produces a legitimate number. For example, the following numbers are only a few of the many that can be made from a roll of a **5**, **2**, × (multiply), ÷ (divide), and − (subtract): 5; 2; 25; 52; 10 (2×5); 32 (2^5); $\frac{2}{5}$ ($2 \div 5$); $2\frac{1}{2}$ ($5 \div 2$); 3 ($5 - 2$); −3 ($2 - 5$); 2.5 ($5 \div 2$); and 0.4 ($2 \div 5$).

4. After three minutes the roller yells "stop!" Every player must stop writing numbers and count how many different numbers they wrote.

🎲 Winning

The player who creates the most correct numbers wins.

🎲 Playing Variations

➤ Play three games. The player who makes the most numbers in any one of the three games is the grand winner.

➤ Set a different time limit. Five minutes works well for experienced players.

➤ Change the operations on the cubes. For younger players use only addition and subtraction signs.

➤ If two players work together, one of them can create numbers and announce how they were constructed while the other records them and the equations that produced them.

➤ Have each player write down both the numbers they create and the equations that produced them.

➤ One person can play Numbers as a solitaire game. The person plays three rounds a day for several weeks, records the high score for each day, and tries to keep beating the previous day's score.

What's Up?

SKILL AREAS:
division with remainders,
divisibility rules

Object: Players cross out a number between 10 and 81 and then roll a numbered cube. The number rolled is divided into the crossed-out number. The remainder is the player's score.

Number of Players: two to eight

Materials: two cubes, lined paper, and pencils

Preparation: On the faces of one cube write the numbers 1, 3, 5, 7, 7, and 9. On the faces of the other cube write the numbers 1, 2, 4, 6, 6, and 8. Write the numbers from 10 to 81 on a sheet of lined paper, leaving about two inches between columns of numbers, as shown. Photocopies of the playing board on the next page (which is a modification of a 0 to 99 number chart) can also be used.

10	28	46	64
11	29	47	65
12	30	48	66
13	31	49	67
14	32	50	68
15	33	51	69
16	34	52	70
17	35	53	71
18	36	54	72
19	37	55	73
20	38	56	74
21	39	57	75
22	40	58	76
23	41	59	77
24	42	60	78
25	43	61	79
26	44	62	80
27	45	63	81

Playing

1. Players take turns, rotating clockwise.

2. To start the game, a player crosses out any unused number on the numbered list. All players use the same numbered list. He or she then chooses one of the two cubes, and rolls it.

3. Next, the player divides the number on the cube into the number he or she just crossed out and finds the remainder.

4. The remainder for the division problem is the player's score for that round. Players keep a cumulative sum of their scores from round to round.

5. A player who notices another player make a mistake acquires that player's score for the turn.

6. The game ends when all of the numbers on the number list have been crossed out.

Winning

The player with the largest cumulative score at the end of the game wins.

Playing Variations

➤ To shorten the game, use fewer numbers on the list. Try the numbers between 20 and 60.

➤ Put numbers larger than 100 on the list to provide practice in long division. For example, try those between 580 and 620. If two-digit numbers, such as 23, are written on the cubes, put still larger numbers, such as those between 7250 and 7275, on the list.

What's Up?

10	11	12	13	14	15	16	17	18	19
20	21	22	23	24	25	26	27	28	29
30	31	32	33	34	35	36	37	38	39
40	41	42	43	44	45	46	47	48	49
50	51	52	53	54	55	56	57	58	59
60	61	62	63	64	65	66	67	68	69
70	71	72	73	74	75	76	77	78	79
80	81								

What's Up?

10	11	12	13	14	15	16	17	18	19
20	21	22	23	24	25	26	27	28	29
30	31	32	33	34	35	36	37	38	39
40	41	42	43	44	45	46	47	48	49
50	51	52	53	54	55	56	57	58	59
60	61	62	63	64	65	66	67	68	69
70	71	72	73	74	75	76	77	78	79
80	81								

Tongue-Depressor Games

Tongue depressors, for the most part, are considered a tool useful only to doctors. However, they are also useful to teachers, for they are one of the sturdiest, cheapest, and most easily stored materials for mathematics games. They can be written on, dealt like cards, moved around like dominoes, tossed into the air to land like pick-up-sticks, tied together in various quantities, or cast as randomizing devices like dice.

Tongue depressors can be bought in drugstores in either the adult size (6 by ¾ inches or 15.2 by 1.9 cm) or the child size (5½ by ⅝ inches or 14 by 1.6 cm). In a pinch, ice cream sticks can be used.

Risk It

Object: Similar to Black Jack or Twenty-One. Players attempt to stop picking numbered tongue depressors from a face-down pile just before the cumulative sum of their numbers reaches 21.

Number of Players: two or three

Materials: 25 tongue depressors

Preparation: Mark each tongue depressor on one side, as shown, with one of the following numbers: 0, 0, 0, 1, 1, 1, 1, 1, 2, 2, 2, 2, 3, 3, 3, 3, 4, 4, 4, 5, 5, 6, 7, 8, 9.

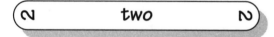

✕ Playing

1. Each player has only one turn. During a turn, the player places all the tongue depressors facedown and mixes them up.

2. The player then turns over tongue depressors one by one and cumulatively adds their numbers. The object is to get a sum close to, but not more than, 20.

3. A player may stop at any point and declare his or her score. The player then turns all the depressors face down and mixes them up for the next player.

4. Players calculate their scores as follows. If the sum of the numbers is less than 20, their score is the cumulative sum. If the sum is equal to 20, their score is 30 (20 plus a bonus of 10). If the sum is more than 20, their score is 0.

✕ Winning

The player with the highest score wins.

✕ Playing Variation

➤ The game can end after each player has five turns. The player with the highest cumulative score wins.

✕ Skill Variations

➤ **Multiplication 1:** Choose any multiplier and have players multiply each tongue depressor they turn over by that multiplier. The number that players try to reach is the multiplier used times 20.

➤ **Multiplication 2:** Use two sets of 20 tongue depressors each (40 altogether). On the backs of the tongue depressors in one set draw a star; on the backs of the others draw a circle. On the fronts of both sets write the following numbers: 0, 0, 1, 1, 2, 2, 3, 3, 4, 4, 5, 5, 6, 6, 7, 7, 8, 8, 9, 9. In this game each player turns over two depressors at a time, one star depressor and one circle depressor; multiplies the numbers; and adds the product to his or her cumulative sum. Scoring: If a player's cumulative sum of products is less than 150, the score is that sum; if equal to 150, the score is 200; if more than 150, the score is zero.

➤ **Fractions:** Write the following numbers on 25 tongue depressors: 0, ½, ½, ½, ½, 1, 1, 1, 1, 1½, 1½, 1½, 1½, 2, 2, 2, 2½, 2½, 2½, 3, 3, 3½, 3½, 3½, 4. Scoring: If the cumulative sum is less than 10, the score is that sum; if equal to 10, the score is 15; if more than 10 the score is 0. Other fractions, such as multiples of ¼, may also be used.

Pick

SKILL AREAS:
addition facts,
column addition

Object: Players receive five numbered tongue depressors. They take turns picking from one another's hands. After four rounds of picking, the player whose sum is the largest wins.

Number of Players: two to five

Materials: 30 tongue depressors

Preparation: Mark three sets of 10 tongue depressors with each of the following numbers: 1, 2, 3, 4, 5, 6, 7, 8, 9, and 10. Write on only one side.

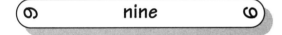

✄ Playing

1. Lay the tongue depressors face down and mix them up.

2. Each player draws five tongue depressors and holds them fan-style to conceal their numbers from other players.

3. The game is played as a series of four rounds.

4. During each round, every player picks one tongue depressor from the hand of the person to the right. The players lay their new tongue depressors face down in front of them until every player has picked. Then all players place their new tongue depressor in their hands and a new round begins.

5. After four rounds, each player adds the numbers in his or her hand.

✄ Winning

The player with the largest sum wins.

✄ Skill Variations

➤ **Addition:** Number the tongue depressors from 40 to 69.

➤ **Multiple Arithmetic Operations:** End the game by having players combine the numbers in their hands as follows: First, each player adds the numbers on any two depressors, then subtracts the number on a third depressor from that sum. Next the player multiplies the number on a fourth depressor by this difference and divides the resulting product by the last number. The player who creates the largest number wins.

Right On

Object: Players discard numbered tongue depressors, attempting to bring the sum of those in a discard pile to exactly 20. The player who does this most often wins the game.

Number of Players: two or three

Materials: 40 tongue depressors

Preparation: Mark 40 tongue depressors, four of each of the following numbers: 1, 2, 3, 4, 5, 6, 7, 8, 9, and 10. Write on and only one side of each tongue depressor.

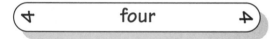

✕ Playing

1. Place the tongue depressors face down between the players and mix them up. Each player picks seven depressors and holds them so that opponents cannot see their numbers. The unpicked depressors become a drawing pile.

2. The first player discards a tongue depressor, face up, to start a discard pile, and calls out its number. The next player discards a second depressor face up onto the discard pile, adds the two numbers together, and calls out the new sum of the discard pile.

3. One after the other, each player discards one tongue depressor and calls out the cumulative sum of all the numbers thus far discarded. The object is to bring the discard pile's sum to exactly 20, and to get rid of all of their depressors.

4. The cumulative sum of the numbers in the discard pile cannot exceed 20. If a player cannot discard a tongue depressor without making the sum larger than 20, he or she must pick one depressor at a time from the drawing pile until one is acquired that can be discarded.

5. The player whose discard brings the sum to exactly 20 gets one point. This player removes the completed discard pile and starts a new one by discarding another depressor.

6. When a player gets rid of all his or her tongue depressors, the game ends and that player receives three points. Otherwise the game ends when no player can discard a depressor and none are left in the drawing pile.

✕ Winning

The player with the most points wins. Ties are possible.

✕ Skill Variation

➤ **Fraction Addition:** To give practice in fraction addition, write fractions on the tongue depressors. To make a game based on multiples of ½, mark two depressors each with the numbers from ½ to 10 in increments of ½. To make a game based on multiples of ¼, mark one depressor each with of the numbers from ¼ to 10 in increments of ¼. Games based on other fractions can be made in a similar way.

Flip

SKILL AREAS: multiplication, addition

Object: Players flip nine numbered tongue depressors into the air. The numbers that land face up are multiplied together. The player with the largest product wins.

Number of Players: two to six

Materials: nine tongue depressors, paper, and pencils

Preparation: Mark one side of each tongue depressor with a number from 1 to 9.

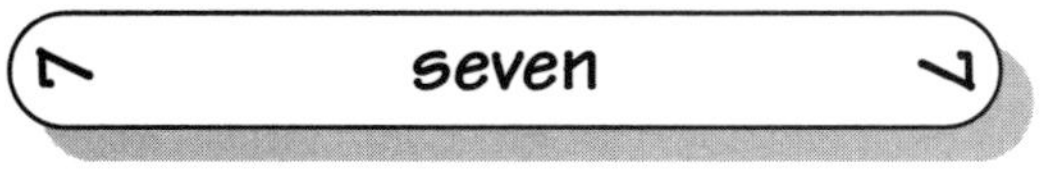

Playing

1. There are 10 rounds. Each player takes one turn during each round.
2. To begin a turn, a player flips the nine tongue depressors into the air one by one. (They can be flipped the same way a coin is flipped.)
3. During a turn a player multiplies together all of the numbers that land face up. For example, if a 3, 4, 5, 6 and 9 land facing up the player would multiply $3 \times 4 \times 5 \times 6 \times 9$ and get a product of 3240. This is the player's score for the turn.
4. A player completes a turn by adding the score just made to the sum of scores he or she made in previous turns. The next player then takes a turn. (Players should carefully check each other's multiplication and addition calculations. Make calculators available to settle disputes.)
5. After 10 rounds, the game ends.

Winning

The player with the largest total score wins.

Playing Variations

- At the end of each round award one point to the player with the largest product. The player with the most points wins.
- Any player who gets a product larger than 2500 gets one point. The player with the most points after 10 rounds wins.
- Reduce the number of rounds.

Skill Variation

- **Addition:** Have players add the numbers on the face-up tongue depressors.

Slide

Object: Similar to the Pea Under the Thimble game. Three tongue depressors whose numbers a player has just viewed are placed face down and mixed up. The player then points to two and guesses their product.

Number of Players: two to six

Materials: nine tongue depressors

Preparation: Mark the tongue depressors on one side with the numbers from 1 to 9.

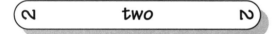

✄ Playing

1. Choose a dealer.

2. The game is played in rounds.

3. To begin a round the dealer turns all nine tongue depressors face down and mixes them. Then the dealer randomly chooses three tongue depressors and turns them face up so that all players can see their numbers. The other six tongue depressors are pushed aside.

4. Next, the dealer turns the three chosen tongue depressors face down in front of the players and quickly mixes them by sliding them around using both hands. The dealer then lines up the three chosen tongue depressors in a row and asks the players for bets. Each player bets by pointing to two tongue depressors and guessing

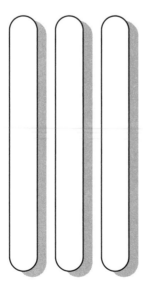

the product of their numbers or have them write their bets down by recording which depressors they are betting on and what the product is. For example, they might write: (left TD) × (middle TD) = 36.

5. To end a round the dealer turns the three chosen tongue depressors face up, examines players' bets, and each player who bet correctly scores one point. A new round then begins.

6. Players keep track of their cumulative scores.

7. The game ends either after a designated period of time or when a player makes a total score of ten points.

✄ Winning

The player with the highest score at the end of the game wins.

✄ Skill Variation

➤ **Addition:** Players guess the sum of the numbers on the tongue depressors, rather than the product.

Togol

Object: Players take turns guessing which of eight numbered tongue depressors lying face down is the largest. Points are awarded for accurately guessing and ordering numbers.

Number of Players: two

Materials: 20 tongue depressors and a small box that will hold them

Preparation: Number the tongue depressors from 1 to 20. Use both written and numeric forms. Write on one side only.

✕ Playing

1. There are ten rounds in the game. During each round one player is the picker and the other a guesser. Players switch roles after each round.

2. To start a round, the picker places all of the tongue depressors in the box, then secretly picks eight of them from the box and places them face down on the table. This is done without letting the guesser see what the numbers are.

3. During the round the guesser turns the tongue depressors face up one at a time. The guesser attempts to stop turning over tongue depressors when the one with the largest of the eight numbers is turned face up. The guesser cannot go back to a previously turned-up depressor; the decision to turn over another depressor is final, even though it is made without knowing the number on it. If the guesser turns over all eight depressors, the last one is considered to be the chosen one.

4. To end a round, the guesser turns all eight tongue depressors face up and orders them from smallest to largest.

5. Score each round as follows:

   ➤ three points to the guesser if he or she chooses the largest number

   ➤ one point to the guesser if he or she correctly arranges the tongue depressors in order from smallest to largest

   ➤ one point to the picker if the guesser does not choose the largest number.

✕ Winning

The player with the highest cumulative score after 10 rounds wins.

✕ Playing Variations

➤ A simpler scoring system awards only one point to the guesser if the guesser chooses the largest number and correctly arranges the tongue depressors from smallest to largest.

➤ Other numbers can be used. For example, the numbers from 1 to 100 can be written on a set of 100 tongue depressors.

Guess It

Object: Two players take turns picking numbered tongue depressors and letting each other guess if they contain an odd or even number. Players acquire tongue depressors if they guess correctly. The player who acquires the most, wins.

Number of Players: two

Materials: 100 tongue depressors

Preparation: Number the tongue depressors from 1 to 100. Use both written and numeric forms. Write on only one side.

 twenty-seven

✳ Playing

1. Turn the tongue depressors face down.

2. The first player picks up a tongue depressor, keeping its blank side toward the second player. The second player guesses whether it has an odd or even number on it.

3. If the second player guesses correctly, he or she takes the tongue depressor. Otherwise, the first player gets to keep it.

4. The players then switch roles—the second player picks up a tongue depressor and the first player guesses whether the number on it is odd or even.

5. Players continue alternating until there are no more depressors left face down. Each then counts the depressors acquired.

✳ Winning

The player with the most tongue depressors wins.

✳ Playing Variation

➤ Use 30 tongue depressors numbered 1 to 10; 38 numbered 1 to 19; or 50 numbered 1 to 50.

✳ Skill Variations

➤ **Inequalities 1:** Players guess whether the number is greater than 50, equal to 50, or less than 50.

➤ **Inequalities 2:** The player who picks up a tongue depressor also chooses any whole number, X, between 1 and 100 and asks: "Is the number smaller than X, larger than X, or equal to X?"

➤ **Multiples/Factors:** Have the player who picks up the tongue depressor ask any of these questions: "Is the number a multiple of X?" "Does this number have a factor of X?" or "Is this number prime or composite?" (X would be a whole number between 2 and 10.)

➤ **Metrics:** Write 100 different metric lengths on the tongue depressors, half of them greater than one meter and half of them less than one meter. Players ask: "Is the length greater than or less than one meter?" Similar games emphasizing other metric units may be made.

Get Seven

Object: Each player receives seven tongue depressors marked with division problems. Players pass one depressor to the player on their left and receive one from the player on their right. The first player to get seven depressors with the same quotient wins.

Number of Players: four to six

Materials: 48 tongue depressors

Preparation: Write the following problems on one side of each tongue depressor.

$1\overline{)4}$ $2\overline{)8}$ $3\overline{)12}$ $4\overline{)16}$ $5\overline{)20}$ $6\overline{)24}$

$1\overline{)5}$ $2\overline{)10}$ $3\overline{)15}$ $4\overline{)20}$ $5\overline{)25}$ $6\overline{)30}$

$1\overline{)6}$ $2\overline{)12}$ $3\overline{)18}$ $4\overline{)24}$ $5\overline{)30}$ $6\overline{)36}$

$1\overline{)7}$ $2\overline{)14}$ $3\overline{)21}$ $4\overline{)28}$ $5\overline{)35}$ $6\overline{)42}$

$7\overline{)28}$ $8\overline{)32}$ $9\overline{)36}$ $10\overline{)40}$ $11\overline{)44}$ $12\overline{)48}$

$7\overline{)35}$ $8\overline{)40}$ $9\overline{)45}$ $10\overline{)50}$ $11\overline{)55}$ $12\overline{)60}$

$7\overline{)42}$ $8\overline{)48}$ $9\overline{)54}$ $10\overline{)60}$ $11\overline{)66}$ $12\overline{)72}$

$7\overline{)49}$ $8\overline{)56}$ $9\overline{)63}$ $10\overline{)70}$ $11\overline{)77}$ $12\overline{)84}$

$6\overline{)30}$

Playing

1. Lay the tongue depressors face down.
2. Each player draws seven tongue depressors, hiding the problems on them from the other players. The remaining depressors are removed, except for one which remains face down.

3. The first player chooses one tongue depressor he does not want. He places it face down in front of the player to his left, and then picks up the extra depressor left on the table.

4. Play now rotates clockwise. Each player in turn passes an unwanted tongue depressor face down to the player on his or her left and then picks up the one given by the player on the right. The objective is to get seven tongue depressors with the same quotient.

5. The game continues in this manner, with players always placing one tongue depressor on the table before picking up the next one.

Winning

The first player to get seven tongue depressors with the same quotient wins.

Playing Variation

➤ Players sit in a circle. Simultaneously, everyone gives a tongue depressor to the person on their left and immediately picks up the depressor received.

Skill Variations

➤ **Division/Fractions/Decimals/Percents/ Multiple Arithmetic Operations:** Mark sets of 48 tongue depressors to drill other arithmetic facts: other division quotients; division remainders (38 ÷ 6 will then have a value of 2); equivalent forms of fractions, decimals, and percents (½, .5, and 50% will be equivalent); or different ways of writing the same number (5 + 4, 3 × 3, 12 – 3, and 27 ÷ 3 will be equivalent). For each game construct four sets of twelve equivalent numbers or problems.

Equato

Object: Players race to be the first to use five numbers along with any arithmetic operations they choose, to form a specified number.

Number of Players: two to six

Materials: 50 tongue depressors

Preparation: Write one of each of the following numbers on one side of a tongue depressor, as shown: 1, 1, 1, 2, 2, 2, 3, 3, 3, 4, 4, 4, 5, 5, 5, 6, 6, 6, 7, 7, 7, 8, 8, 8, 9, 9, 9, 10, 10, 10, 11, 11, 12, 12, 13, 13, 14, 14, 15, 15, 16, 17, 18, 19, 20, 21, 22, 23, 24, 25.

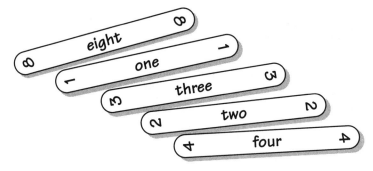

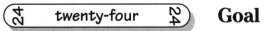 **Goal**

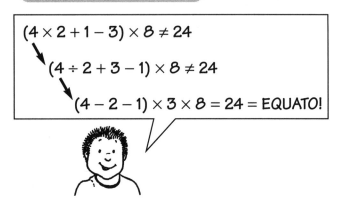

$$(4 \times 2 + 1 - 3) \times 8 \neq 24$$
$$(4 \div 2 + 3 - 1) \times 8 \neq 24$$
$$(4 - 2 - 1) \times 3 \times 8 = 24 = \text{EQUATO!}$$

✂ Playing

1. Equato is played in rounds, during which everyone plays simultaneously.

2. To begin a round the tongue depressors are placed face down and mixed up. Each player picks five tongue depressors and places them face up in front of him or her. A final depressor is then turned face up—its number is the "goal" for the round.

3. The players race to be the first to use the numbers on their five tongue depressors along with any arithmetic operations—addition, subtraction, multiplication, or division—to form a number equal to the goal for the round. Each depressor number must be used exactly once, they may be used in any sequence, and the player may move the depressors around. Any arithmetic operation may be used any number of times in any sequence (see the sample game).

4. The first player to create an equation calls out "Equato" and slaps the goal depressor. Within 30 seconds that player must show the equation created. If it is valid, that player gets one point. If not, the round continues with that player disqualified for the remain-

der of it. If no player can create an equation in a reasonable amount of time (say, three minutes) the round is ended with no score.

5. All the tongue depressors are laid face down and a new round begins.

6. The game ends when one player earns three points.

✂ Winning

The first player to get three points wins.

✂ Playing Variation

➤ Create simpler games by marking the tongue depressors with only the numbers from 1 to 20 (or 1 to 10) and by allowing players to use four (or three) of their five depressors to make the equation.

Ladders

SKILL AREAS:
ordering numbers, comparing numbers, fractions

Object: Players pick six tongue depressors and try to be the first to arrange them in order from smallest to largest.

Number of Players: two to four

Materials: 50 tongue depressors

Preparation: Write one of each of the following fractions on one side of two tongue depressors: 0/2, 1/2, 2/2, 0/4, 1/4, 2/4, 3/4, 4/4, 0/3, 1/3, 2/3, 3/3, 0/12, 1/12, 2/12, 3/12, 4/12, 5/12, 6/12, 7/12, 8/12, 9/12, 10/12, 11/12, 12/12.

Playing

1. Place the tongue depressors face down and mix them up.
2. Each player picks six tongue depressors. Each tongue depressor is turned face up and placed in front of the player just above the last depressor picked. The first depressor picked will lie closest to the player and the last one picked will lie farthest from the player, forming a ladder as shown below. Players may not alter this initial sequencing, except when taking a turn.

3/12 three twelfths 3/12	← sixth pick
1/4 one fourth 1/4	← fifth pick
5/12 five twelfths 5/12	← fourth pick
1/3 one third 1/3	← third pick
1/4 three fourths 1/4	← second pick
1/12 one twelfth 1/12	← first pick

3. Players take turns in clockwise rotation. The players try to replace depressors in their ladders so that the numbers become ordered from smallest to largest, with the smallest number closest to a player. Two equivalent numbers cannot be in a player's ladder.
4. In turn, players pick one depressor from those lying face down and decide whether to discard it or use it. Discarded depressors are placed face up on a discard pile. If the player uses the depressor, he or she exchanges it with one of the depressors in his or her ladder—without disturbing the order of the depressors—and the exchanged depressor is discarded. For example, during the game shown the player picked a 2/3, exchanged it with the 3/12, and discarded the 3/12.

2/3 two thirds 2/3	largest
1/2 one half 1/2	↑
5/12 five twelfths 5/12	**Winning Sequence**
1/3 one third 1/3	
1/4 one fourth 1/4	↓
1/12 one twelfth 1/12	smallest

Winning

The winner is the first player to get all six depressors sequenced from smallest to largest.

Skill Variations

➤ **Signed Numbers:** Write the integers from –25 to +25 on the depressors.

➤ **Other Types of Numbers:** Write the counting numbers from 1 to 50 on the depressors. Create other games using decimals, mixed numbers, or unsolved equations (sequenced according to their answers).

Card Games

Card games have long been a popular means of entertainment for both children and adults. The following games do not use regular playing cards, but are otherwise similar to many established card games.

Index cards can be used to make playing cards. Blank playing cards are sold by educational materials supply companies. Number 68 round-corner tickets, sold by most printers, are also ideal. Cards can also be cut from any stiff paper stock such as oak tag or tag board, to the size of approximately 2.25 by 4 inches (5 by 8 cm). When cutting cards from stiff paper stock, it is best to first draw the cards on the paper stock, laminate the entire piece of paper, and then cut out the cards. Clear contact paper can be used instead of lamination film.

Operations War

SKILL AREAS:
multiplication, inequalities

Object: Similar to War. Two numbered cards are dealt to each player. The numbers on the two cards are multiplied. The player with the larger product wins and takes all four cards.

Number of Players: two

Materials: 40 cards

Preparation: On four sets of 10 cards write each of the numbers from 1 to 10. Write on only one side of each card.

Playing

1. Each player receives 20 cards.

2. Both players place their cards face down, in a neat stack.

3. From the top of their stacks, both players simultaneously deal themselves one card face down—without looking at it—and then deal their opponent one card face up.

4. Both players then simultaneously turn over their face-down cards, multiply the numbers on the two cards together, and announce the product. The player with the larger product keeps all four cards and places them face down on the bottom of his or her stack of cards. (Players should check each other's multiplication.)

5. Play continues repeatedly as described above.

6. If a tie occurs, both players repeat the process of dealing a face down card and a face up card. The winner of this second round takes all eight cards.

Winning

If a player acquires all 40 cards, that player wins. Otherwise, the player with the most cards after some predetermined time is the winner.

Skill Variations

➤ **Addition or Subtraction War:** Have players find either the sum or difference of the numbers on the two cards.

➤ **Three Person Addition/Multiplication War:** Each player gets 13 cards to stack. Have each player deal him- or herself one card face down and then deal each opponent one card face up. Players add the face up cards and then multiply the sum by the face down card. The player with the largest result acquires all nine cards.

Bongo

Object: A variation of Bingo using two sets of cards, one with problems and the other with answers. Five problem cards are dealt to each player. Players match their problems to the answers called.

Number of Players: three to 11, or whole class if several sets are used

Materials: 55 red cards and 55 blue cards (or any other two colors)

Preparation: Write one of these 55 multiplication problems on one side of each red card.

1×5	1×6	1×7	1×8	1×9	1×10
2×5	2×6	2×7	2×8	2×9	2×10
3×5	3×6	3×7	3×8	3×9	3×10
4×5	4×6	4×7	4×8	4×9	4×10
5×5	5×6	5×7	5×8	5×9	5×10
	6×6	6×7	6×8	6×9	6×10
1×4		7×7	7×8	7×9	7×10
2×4	1×3		8×8	8×9	8×10
3×4	2×3	1×2		9×9	9×10
4×4	3×3	2×2	1×1		10×10

Write one of the following 55 products on one side of each blue card.

5	6	7	8	9	10
10	12	14	16	18	20
15	18	21	24	27	30
20	24	28	32	36	40
25	30	35	40	45	50
	36	42	48	54	60
4		49	56	63	70
8	3		64	72	80
12	6	2		81	90
16	9	4	1		100

Playing

1. Decide whether to deal problem cards or answer cards to players. Either can be dealt, but not a mix of both. (These directions assume that problem cards are dealt.)

2. Choose a dealer. The dealer deals five problem cards to each player and sets aside any that are left over.

3. The players place their five problem cards face up in front of themselves.

4. The deck of answer cards is placed face down in front of the dealer, who picks them up one by one, holds each facing the players, and announces the number.

5. All players with a problem card that matches the dealer's answer card turn that card face down. A player may turn over only one problem card for each answer card held up, even if several cards match.

6. The dealer continues turning up answer cards until a player calls "Bongo!" A player calls "Bongo!" as soon as all of his or her cards are turned face down.

7. The player who called "Bongo" announces the problems on his or her cards while the dealer checks that they match the answer cards turned up. If not, play continues.

Winning

The first player to correctly call "Bongo" wins.

Skill Variations

➤ **Addition, Division, Subtraction, Fractions:** Different types of Bongo can be constructed by creating 55 pairs of matching mathematical expressions and writing them on two sets of different colored cards.

Little Shot

Object: As numbered cards are drawn from a deck nine times, players write the resulting numbers in the nine cells of a 3 × 3 grid. Players calculate scores by adding the numbers in the grid's rows, columns, and diagonals.

Number of Players: two to whole class

Materials: six cards, copies of the game sheet for each player, and pencils

Preparation: Write each of the numbers from 1 to 6 on one side of each card. Hand out copies of the game sheet on the next page or have each player make a copy of it.

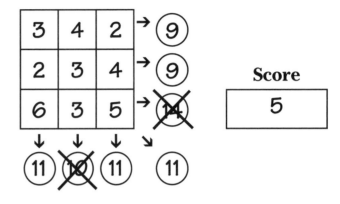

Score

5

Playing

1. Each player gets a game sheet and a pencil.

2. Choose one person to hold, shuffle, and draw the cards, and to announce each number drawn. Cards are drawn nine times during the game. After each card is drawn and its number announced, the card is returned to the deck, and the deck is shuffled.

3. After each number is announced and before the next card is drawn, players write the announced number in any empty square on their game sheet. Once written, a number cannot be moved.

4. After the nine numbers are written, the players find the sums of the numbers across the three rows, down the three columns, and through the diagonal. They record the sums in the corresponding circles.

5. Players cross out any sum that appears in only one circle.

6. The number of sums that are not crossed out is a players score.

Winning

The player with the largest score wins.

Playing Variations

➤ Drill other addition facts by writing different numbers on the cards.

➤ Play three successive games of Little Shot. The winner is the player with the largest cumulative score.

➤ Calculators can be used to check scores.

Little Shot

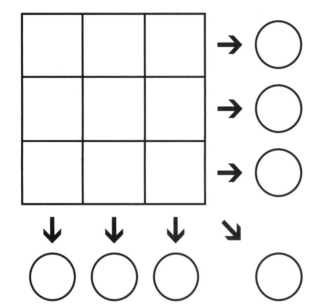

Score

Little Shot

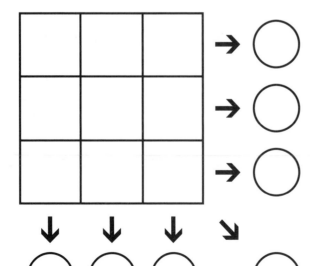

Score

Place-a-Bet

SKILL AREAS:
place value, inequalities

Object: Players secretly construct a three-digit number from three numbered cards. Then they each bet whether their number is the highest, middle, or lowest. A player who bets correctly gets one point.

Number of Players: three

Materials: 40 cards, and copies of the playing sheet on the next page

Preparation: Mark four sets of 10 cards with each of the numbers from 0 to 9. Write only one number on one side of each card. Reproduce copies of the playing sheet on the next page for each player or have each player copy it.

♦ Playing

1. Give each player a playing sheet, choose a dealer, and have the dealer shuffle the cards.

2. There are 10 rounds in a game. To begin each round, deal three cards to each player and proceed as follows.

3. Players look at their cards, without letting their opponents see them, and secretly put them together in any way they wish to form a three-digit number. They record that number in the Number column of their playing sheets.

4. Each player now bets whether his or her number is the highest, middle, or lowest of the three numbers created by the three players. They all record their bets on their playing sheet next to their numbers by marking an X in either the High, Middle, or Low column. They may mark in only one column.

5. When all three players have marked their bets, they turn their cards face up. Each player tells the others, using place value terminology, the number he or she made. Each player shows the others this number written on the playing sheet. Each player reveals his or her bet.

6. A player who makes a correct bet gets one point. A player who makes an incorrect bet gets zero points. Players record 0 or 1 in the Points column of their playing sheets.

7. To end each round, players return their cards to the dealer, who shuffles and redeals them to begin the next round.

8. At the end of 10 rounds, players add their points and record the sum in the Total Score box.

♦ Winning

The player with the highest total score wins.

♦ Playing Variations

➤ Play the game with two-, four-, five-, or six-digit numbers.

➤ Penalize players who read their number wrong during a round by ruling that they may not score a point on that round.

Place-a-Bet

Round	Number	Bets			Points
		High	Middle	Low	
1					
2					
3					
4					
5					
6					
7					
8					
9					
10					
				Total Score	

Place-a-Bet

Round	Number	Bets			Points
		High	Middle	Low	
1					
2					
3					
4					
5					
6					
7					
8					
9					
10					
				Total Score	

Running Sum

Object: Players place numbered cards end-to-end to form a connected network of cards, in a manner similar to Scrabble™. To place cards, card numbers must add up to a predetermined sum.

Number of Players: two to four

Materials: 63 cards

Preparation: Mark seven sets of nine cards with each of the numbers from 1 to 9. Write one number on each card (on one side only).

🂠 Playing

1. To begin, remove two 5's, two 6's, two 7's, two 8's, and two 9's from the deck. Place these 10 cards face down, mix them up, and then pick two of them. The sum of the numbers on these two cards is the running sum for the game.

2. Return the 10 cards to the deck, shuffle the deck, and deal each player 15 cards face down. Players keep the numbers on their cards hidden from opponents until they use them.

3. Players take turns in clockwise rotation. Their objective is to get rid of as many cards as possible.

4. On the first turn of the game, the first player places two or more cards face up on the table. The numbers on the cards must add up to the running sum. The player lays the cards down end-to-end in a straight line (called a run). For example, if the running sum is 16, the first player could play 3, 5, and 8 (see the sample game).

5. During each succeeding turn of the game a player adds a new run of cards to those already on the table. Each new run must be connected to a card in a previous run, as in

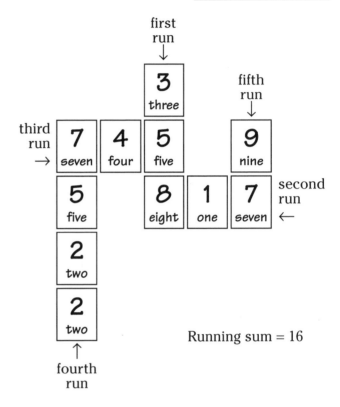

Running sum = 16

Scrabble™. Also, the sum of the new cards plus the interconnecting old card must equal the running sum. For example, note in the sample game that the cards in each run add up to 16 (the running sum in this game).

6. If a player cannot play any cards on a turn, that player must pass, and loses the chance to get rid of cards.

7. The game ends either when one player has no more cards left or when all players have passed and none can play any of their remaining cards.

🂠 Winning

The player who first uses up all 15 cards wins. If the game ends with all players still holding some cards, then the player with the fewest cards wins.

Slap It

Object: Players take turns flipping over face-down cards. When two match, the first player to slap one of them gets both. The player who gets the most wins.

Number of Players: two to four

Materials: 36 cards

Preparation: On one side of each card write one of the following numbers or problems:

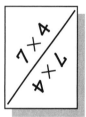

8	2 × 4	4 × 2	1 × 8
12	3 × 4	2 × 6	1 × 12
16	4 × 4	2 × 8	1 × 16
20	5 × 4	2 × 10	1 × 20
24	6 × 4	2 × 12	3 × 8
28	7 × 4	2 × 14	1 × 28
32	8 × 4	2 × 16	1 × 32
36	9 × 4	2 × 18	1 × 36
40	10 × 4	2 × 20	5 × 8

Playing

1. Place all of the cards face down on the playing surface and mix them up.

2. Play rotates clockwise. In turn, each player chooses one card and immediately places it number-side-up on the playing surface. Cards that are turned number-side-up remain that way throughout the game, until they are won by a player and removed from the playing surface.

3. Each time a player turns up a card that has a value that is equal to a card that is already face up, all players try to be the first to slap a hand over the matching card that was previously face up. The player to first slap the card with the equivalent value to the card that was just turned over wins both cards and removes them from the table.

4. The game ends when all of the cards are face up. Players then count the number of cards they have won.

Winning

The player with the most cards wins.

Playing Variation

➤ Have players turn over two cards at the same time. If they match, the player keeps them; otherwise the player turns them face down before play passes to the next person.

Skill Variations

➤ **Money Equivalents:** Put the following money equivalent problems and answers on the cards: penny, nickel, dime, quarter, half-dollar, dollar, 1 dollar, 5 dollars, ¢, 1¢, 5¢, 10¢, 25¢, 50¢, 100¢, 4 quarters, $5.00, cents, $.01, $.05, $.10, $.25, $.50, $1.00, 10 dimes, 10 dollars, 2 quarters, one cent, 5 pennies, ten cents, fifty cents, $, 20 nickels, $10.00, 5 dimes.

➤ **Fractions/Decimals:** Put equivalent fractions and decimals on the cards.

Call It

Object: Players attempt to discard numbered cards by taking turns placing a card onto a discard pile and multiplying its number by the number on the previously discarded card.

Number of Players: two to five

Materials: 40 cards

Preparation: On four sets of 10 cards write one of each of the numbers from 1 to 10. Write on only one side of each card.

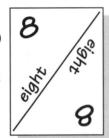

Playing

1. Deal six cards to each player. Deal another card face up to start a discard pile. Place the rest of the cards aside.

2. Players take turns clockwise. Players attempt to get rid of all their cards.

3. On a turn a player places one card onto the discard pile, multiplies its number by the number of the card on the top of the discard pile (the previously played card), and calls out the product.

4. If the product is correct, the player leaves the card on the discard pile and the next player takes a turn. If the product is incorrect, the player must take back the discarded card. The next player then takes a turn.

Winning

The winner is the first player, after each player has had six turns, to discard all of his or her cards. If several players discard all six of their cards within their first six turns, they are all declared winners.

Playing Variations

➤ Play several games, have players take turns going first, and see who wins the most games after each player has had a turn going first.

➤ Players can be required to state aloud the multiplication problem as well as the product.

Skill Variations

➤ **Addition or Subtraction:** Have players find either the sum or difference of the numbers on the cards.

➤ **Fractions:** Addition or subtraction of fractions can be practiced with this game by writing one of each of the following numbers on one side of each card.

0	¼	½	¾
1	1¼	1½	1¾
2	2¼	2½	2¾
3	3¼	3½	3¾
4	4¼	4½	4¾
5	5¼	5½	5¾
6	6¼	6½	6¾
7	7¼	7½	7¾
8	8¼	8½	8¾
9	9¼	9½	9¾

What's Next?

Object: Each player draws an arrangement of place holders that represent an arithmetic problem. As numbered cards are drawn one by one, players fill in the place holders with these numbers and then solve their problems.

Number of Players: three to whole class

Materials: Ten cards, paper, and pencil for each player

Preparation: On one side of each card write a number from 0 to 9. Decide what type of arithmetic problem the game will deal with (see samples below). Draw a picture depicting the form of the problem on a chalkboard or chart paper, and display it for players to copy.

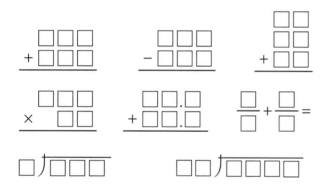

🗂 Playing

1. Choose a person to be a caller. Have each player copy the picture of the arithmetic problem on a sheet of paper.

2. The game is played in rounds with all players playing simultaneously. There are as many rounds as there are empty place holders in the problem.

3. To begin a round, the caller shuffles the cards, holds them so that the numbers on the cards cannot be seen, randomly picks one card, holds it up so that all players can see the number, and announces the number aloud.

4. Each player writes the number announced by the caller in any empty place holder. Once a number is written, it may not be moved.

5. At the end of a round the caller replaces the card in the deck and picks another card.

6. When all of the place holders in the players' problem have been filled, the players solve their own problem. They then check each other's problems and solutions.

🗂 Winning

The player whose problem produces the largest answer wins. There may be ties.

🗂 Playing Variations

➤ Once the caller picks a card, it is not returned to the deck for the rest of the game. This method works only when there are fewer than seven place holders in the problem.

➤ With fewer than six players (or an entire class divided into fewer than six teams), play 10 games and see who wins the most.

➤ The winner can be the player whose problem produces the smallest answer.

➤ Record sheets can be constructed and used for a series of games. Placing six to eight problems on a record sheet works well. Different types of problems can be placed on a single record sheet.

Two Numbers

Object: Cards with two numbers on them are revealed one at a time. Players calculate a card's value by adding, subtracting, multiplying, or dividing its numbers. If two cards have the same value, players can capture them.

Number of Players: three to six

Materials: 75 cards

Preparation: Mark one side of each card with one of the pairs of numbers listed below, as shown in the diagram.

²	⁴
4	2

1, 1	2, 2	3, 3	4, 4	5, 5	6, 6	7, 7
1, 2	2, 3	3, 4	4, 5	5, 6	6, 7	7, 8
1, 3	2, 4	3, 5	4, 6	5, 7	6, 8	7, 9
1, 4	2, 5	3, 6	4, 7	5, 8	6, 9	7, 10
1, 5	2, 6	3, 7	4, 8	5, 9	6, 10	
1, 6	2, 7	3, 8	4, 9	5, 10	6, 12	8, 8
1, 7	2, 8	3, 9	4, 10	5, 15	6, 18	8, 9
1, 8	2, 9	3, 10	4, 12	5, 20	6, 24	8, 10
1, 9	2, 10	3, 12	4, 16	5, 25	6, 30	
1, 10	2, 12	3, 15	4, 20	5, 30		9, 9
	2, 14	3, 18	4, 24			9, 10
	2, 16	3, 21				
	2, 18					10, 10

🃏 Playing

1. A card's values are calculated by adding, subtracting, multiplying, or dividing its two numbers. For example, the card with 2 and 4 on it has the following values: 6 (2 + 4); 2 (4 − 2) or (4 ÷ 2); 8 (4 × 2); ½ (2 ÷ 4); −4 (2 − 4). (Note: Values can be limited to positive whole numbers.)

2. The dealer shuffles the cards and places the deck face down. The dealer then deals the top card face up. This card is called the object card. Players calculate its values.

3. The dealer now turns over the cards from the top of the deck and places them face up in a heap called a playing pile. The interval of time between dealing each card can very from five to forty seconds, depending upon the players' skills.

4. As each card lands face up on the playing pile, the players calculate its values. If any of these values equal a value of the object card, the players try to slap the object card.

5. The first player to slap the object card captures both it and all the cards that have accumulated in the playing pile. A player who erroneously slaps the object card is eliminated from the game until another player slaps it.

6. As soon as a player captures the object card and the playing pile, that player takes them. The dealer deals a new object card and begins dealing cards onto a new playing pile.

7. When the dealer has dealt all of the cards in the deck, players count up the number of cards they captured, and record this number. Then a new player becomes dealer.

8. The dealer does not capture cards while dealing. During the game the deal rotates clockwise. After all players have had a turn dealing, the game ends. Players then calculate the number of cards they captured during the entire game.

🃏 Winning

The player who captures the most cards wins.

🃏 Playing Variation

➤ For easier games have the dealer choose only one of the values of the first card turned over to be the objective.

Mathino

Object: Similar to Casino. Players acquire numbered cards by constructing equations.

Number of Players: two to four

Materials: 60 cards

Preparation: Mark five sets of 12 cards with each set having the numbers from 1 to 12. Mark cards on only one side.

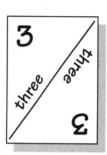

Playing

1. Mathino is played in rounds. To start the first round, the dealer deals four cards face down to each player and four cards face up onto the table to make a "field". In successive rounds four cards are dealt face down to each player. Players place cards they acquire in a pile in front of them. A round ends when each set of four cards is played.

2. During each round players pick up the cards dealt, concealing their values from others. Players take turns, with play rotating clockwise. On a turn a player plays only one card and can do one of the following:

> **Match:** If a player's card is the same as a field card, the player can match them on the field and acquire both.

> **Combine:** If a player's card equals the sum of two or more field cards, the player may combine them on the field and acquire them all. For example, a 7 may acquire a combination of a 4 and 3. Several combinations may be acquired simultaneously.

> **Hold:** If several of a player's cards are the same as a single field card or sum of field cards, the player may play one of the cards onto its match or combination, say "Holding (naming the cards)," leave them on the field, and acquire them later in the round. (For example, a player holding two 3's can place one of the 3's on a field 3, say "Holding threes," and use the other 3 to acquire the pile later.)

> **Build:** A player's card may be combined with one or more field cards to build an equation that equals another card in the player's hand. The player puts the cards in a single pile on the field and says "Building (naming the sum)." (For example, a 3 from a player's hand can be put into a pile with a 5 and 2 from the field with the declaration "Building 10 = 3 + 5 + 2.")

> **Steal:** A player may match and acquire any pile of cards that an opponent is holding or building, if that player has the necessary card to do so.

> **Rebuild:** A player may build on an opponent's "build" or on his own "build." (For example, if a player is "building" 6 with 2 and 4, another player may add a 3 and say "Building 9 = 3 + 2 + 4.")

> **Discard:** If a player cannot match, combine, hold, build, steal, or rebuild, the player must add a card to the field.

3. A pile of cards being "held" cannot be built upon. A player does not have to acquire a hold or build immediately.

4. In the final round of the game, deal any remaining cards face up onto the field.

Winning

The player who acquires the most cards wins.

Skill Variation

> **Multiple Operations:** Permit players to subtract, multiply, or divide when they combine.

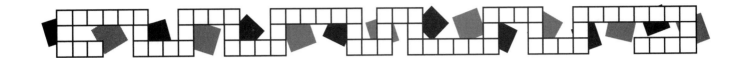

Board Games

Board games have been popular since the time of the Ancient Egyptians. Some involve several players playing together around a single game board. Others involve several players each using a similar game board seated around a "bank."

Thick paperboard is an ideal material to use for board games. A game board can be drawn directly on the paperboard or can be photocopied and pasted to it. Photocopying can enlarge a board to almost any size desired. Decorate the game boards with colored crayons, pencils, or felt-tip markers to highlight a theme. Instructions can be pasted to the back. Game boards can be laminated or covered with clear contact paper to protect them. This also allows a game board to be written on and erased, if appropriate writing implements are used. Dry erase markers and wax crayons work well. Constructing a game board on the inside of a large flat cardboard box (such as a shirt box) is useful because the playing pieces can be stored in the box.

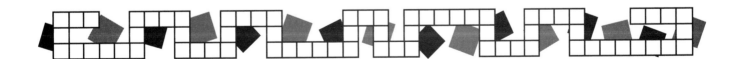

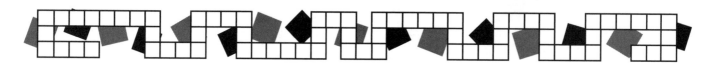

The Great Trading Game

SKILL AREAS:
place value, addition facts,
addition with carrying, money

Object: Players roll dice to determine how many pennies they receive in a race to be the first to get one dollar. Trades for dimes and dollars take place as the game progresses.

Number of Players: two to four

Materials: paper or paperboard (about 8.5 by 11 inches), two dice, 50 counters each of three different colors (or play money), and a box to hold all of the counters

Preparation: Construct a playing board for each player. The next page can be photocopied and cut in half to obtain two game boards.

▦ Playing

1. Players sit around the box of counters. Players agree on which counters will represent pennies, which will be dimes, and which will be dollars.

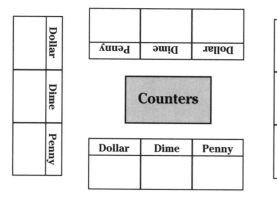

2. Players take turns, rotating play in a clockwise direction.

3. During a turn a player rolls the dice, adds the numbers, and takes that number of pennies. They are placed on his or her game board under the Penny column.

4. As soon as the player collects ten pennies, he or she must trade them in for a dime, if possible. The same applies for trading in dimes for a dollar. This is called *legalizing*. If a player does not legalize when it is possible, the opponent who notices wins one penny from the player who did not legalize.

▦ Winning

The first player to earn a dollar wins the game.

▦ Playing Variations

➤ One player can be the banker. The banker checks all sums, gives out chips, and exchanges coins during trading transactions.

➤ Use wood cubes instead of dice. A cube might be numbered with 3, 4, 5, 6, 7, and 8 to give children practice adding these numbers.

➤ The game board can be constructed to contain an area for Ten Dollars.

➤ Use base-ten blocks instead of counters.

➤ Players can be asked to record equations that represent their progress during the game (previous score + new pennies acquired = new score) on a sheet of paper.

▦ Skill Variations

➤ **Subtraction:** Players start with two dollars on their boards and remove a number of pennies equal to the sum of the numbers rolled on the dice. The first player with no money wins.

➤ **Decimals:** Players record equations corresponding to their moves in decimal notation.

The Great Trading Game

Dollar	Dime	Penny

The Great Trading Game

Dollar	Dime	Penny

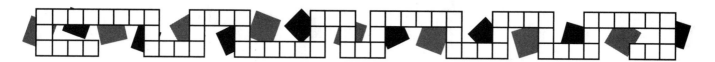

0 to 99 Guess

SKILL AREAS:
inequalities, number patterns, counting, logic

Object: A leader secretly circles one number on the game board. Players try to guess the number by asking yes-or-no questions.

Number of Players: three to five

Materials: paper and crayons

Preparation: Make five photocopies of the game board and laminate them. For sturdier game boards, copy the 100 number chart on paperboard. For disposable game boards, simply photocopy.

▦ Playing

1. The game has a leader, who hides a number on the game board and players who try to guess the hidden number. The game is played in rounds. Players take turns being the leader in successive rounds.

2. To begin a round, the leader circles one number on his or her game board, and covers it or turns it over so players cannot see the circled number.

3. During a round, players take turns, rotating clockwise, asking the leader questions about the number that can be answered with either "yes" or "no," such as, "Is the number larger than 25?" or "Is the number divisible by 5?" During a turn a player may ask only one question. After a question is answered, all players can mark numbers on their game board. For example, if the number is less than 25, players might cross out all numbers larger than 24.

4. A round ends when someone guesses the hidden number. A player can only guess on his or her own turn. If a player determines what the hidden number is as a result of asking a question, the player must wait until his or her next turn to guess the number.

5. The player who guesses the hidden number correctly ends the round and gets one point. Players clean their game boards and keep track of their points.

▦ Winning

The player with the largest number of points after a specified number of rounds or playing time wins.

▦ Playing Variations

➤ To play with a large group of children, divide the group into two teams.

➤ For experienced players, do not use a game board. Have players visualize the game board and write the answers on paper.

➤ Instead of orally answering questions, the leader can write the answers. For example, to the question "Is the number larger than 25?" the leader might write "X > 25".

➤ In a two-person game, players take turns being leader. A player receives a score equal to the number of guesses it took to determine the hidden number, and the player with the smallest cumulative score wins.

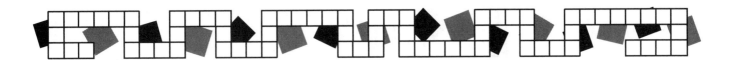

0 to 99 Guess

0	1	2	3	4	5	6	7	8	9
10	11	12	13	14	15	16	17	18	19
20	21	22	23	24	25	26	27	28	29
30	31	32	33	34	35	36	37	38	39
40	41	42	43	44	45	46	47	48	49
50	51	52	53	54	55	56	57	58	59
60	61	62	63	64	65	66	67	68	69
70	71	72	73	74	75	76	77	78	79
80	81	82	83	84	85	86	87	88	89
90	91	92	93	94	95	96	97	98	99

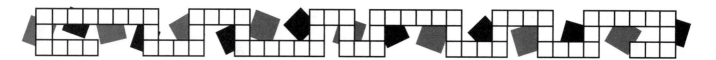

In the Basket

Object: Players roll dice and occupy positions according to the sum of the numbers rolled. Rolling a seven results in a penalty. Rolling a sum occupied by another player results in a reward.

Number of Players: two to five

Materials: large sheet of paperboard, 50 counters, and two dice

Preparation: Copy the following game board on paperboard. If desired, embellish the game board with colors and transform the basket into something of interest to the children playing the game, such as a dinosaur. (The game board can be photocopied.)

🎫 Playing

1. To begin, give each player 10 counters and determine the order in which players will take their turns.

2. In turn, the players roll the two dice and calculate the sum of the numbers rolled. If that circle on the game board with that sum is not occupied by a counter, the player places one counter on that circle and the player's turn ends. If the circle is occupied by a counter, the player takes the counter off of the game board, adds it to his or her counters, and takes another turn. The player continues rolling the dice until he or she can place a counter on an empty circle on the game board. If a player rolls a sum of 7 on the dice, the player deposits one counter in the basket and the player's turn ends.

3. When a player runs out of counters, he or she is out of the game.

4. Play continues until only one player remains.

🎫 Winning

The last player remaining wins.

🎫 Playing Variations

➤ For a quicker game, give players only six counters at the beginning of the game.

➤ If several games will be played, the winner gets a score equal to the number of counters possessed at the end of the game. The "grand winner" of several games is the player with the largest cumulative score.

➤ Players can record on paper or verbally state the equations that correspond to the numbers rolled on the dice. If an equation is incorrect, the player gives the basket one counter and the player's turn ends.

🎫 Skill Variation

➤ **Addition of Three Numbers:** The game can be extended to larger numbers by using three dice. In this case, the game board should contain circles with the numbers 3 to 18. Both 9 and 12 are penalty numbers that require players to deposit one counter in the basket.

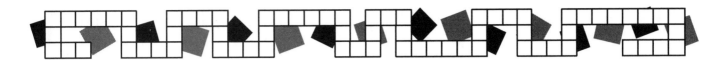

In the Basket

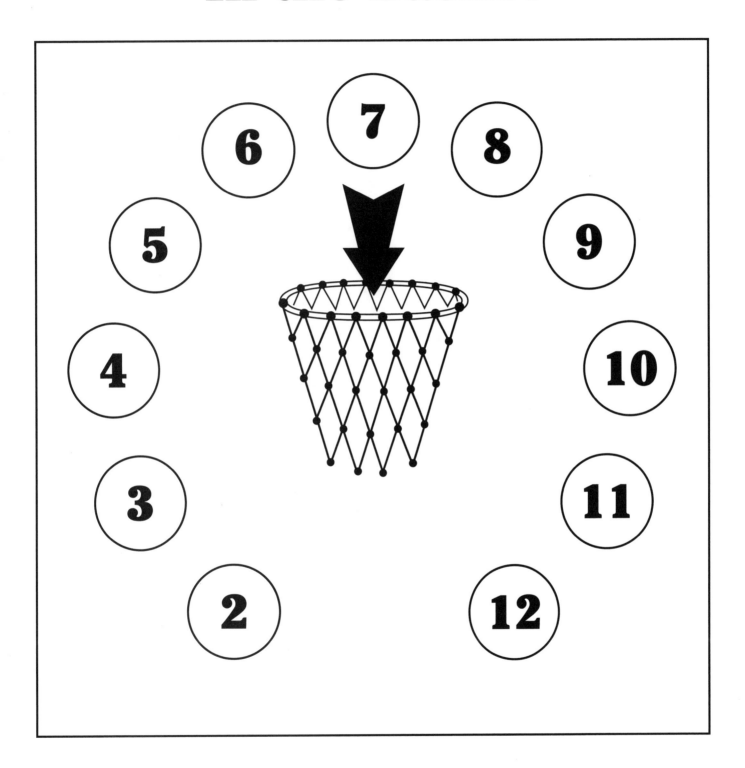

Race to the Nineties

Object: Players pick cards that contain vectors that designate moves as they race from 0 to any number between 90 and 99.

Number of Players: two to four

Materials: paperboard, four different small objects for playing pieces, and 44 blank cards

Preparation: Make and decorate the game board. It can be photocopied. Color the top row (Start) on the board light green and the bottom row (90–99) light blue (numbers and words must show through the coloring). On one side of four sets of 11 cards, draw vectors (arrows), as shown. On the top of each card draw a green line and on the bottom draw a blue line (use the same colors as on the game board to help give orientation to the cards).

🖿 Playing

1. Each player chooses a playing piece and places it on the green bar at the top of the game board. Shuffle the cards and place them in a neat deck face down next to the game board.

2. To start, players move their pieces to 0 and follow directions from 0.

3. Players take turns. During a turn a player picks a card from the deck, turns it arrow side up, and follows the instruction given by the arrow(s) on the card. The arrow pointing straight down means move the playing piece

down one row (which is the same as adding 10 to the number). The arrow pointing diagonally down and to the right means move the playing piece down one row and to the right one space (which is the same as adding 11). After a player has moved, the player's card is placed face down on the bottom of the deck.

4. If a playing piece is moved onto a space already occupied by another playing piece, the playing piece that was on the space first is bumped to the green bar at the top of the board. It starts over from 0.

5. When a card designates a move that will take the player's piece off of the game board, he or she picks a second card. If the second card takes the player off of the board as well, the player stays put and loses his or her turn.

6. During a turn players must state (and/or write) the equation that correspond to their moves. For example, if a player's piece resides on 25 and the player picks a card with an arrow pointing straight up, the player would say " 25 − 10 = 15." If all the equations are written down in a neat row, a player's progress can be easily reviewed.

🖿 Winning

The winner is the first player to reach the bottom row of the game board (any number from 90 to 99).

🖿 Playing Variation

➤ Players can simply state their starting and ending positions. For example, " I started at 25 and ended at 15."

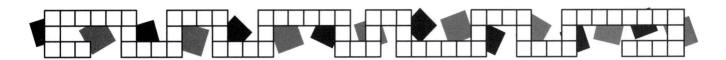

Race to the Nineties

Start									
0	1	2	3	4	5	6	7	8	9
10	11	12	13	14	15	16	17	18	19
20	21	22	23	24	25	26	27	28	29
30	31	32	33	34	35	36	37	38	39
40	41	42	43	44	45	46	47	48	49
50	51	52	53	54	55	56	57	58	59
60	61	62	63	64	65	66	67	68	69
70	71	72	73	74	75	76	77	78	79
80	81	82	83	84	85	86	87	88	89
90	91	92	93	94	95	96	97	98	99

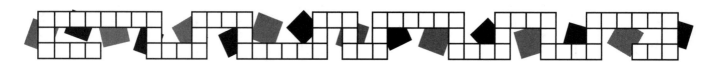

Checker Math

Object: A checkers-like game in which the goal is to get the highest cumulative score rather than to take all the opponent's checkers.

Number of Players: two

Materials: checkerboard, set of 24 checkers, and markers with which to write on the checkerboard

Preparation: On the lighter-colored squares of the checker board, write numbers with the marker as shown.

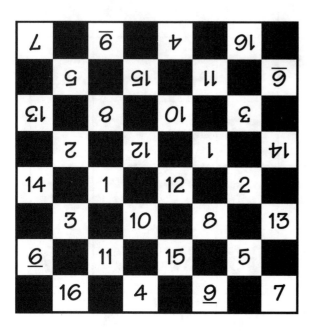

▦ Playing

1. Players sit opposite each other with the board positioned between them. One player has 12 dark checkers and the other has 12 light checkers. Each player puts his or her checkers on the numbered squares in the three rows of squares closest to themselves, as in regular checkers.

2. Players take turns. On a turn a player moves or jumps with one of his or her own checkers. Checkers are moved or jumped on only numbered squares. A move consists of displacing a checker one space forward in a diagonal direction toward the opponents side of the board onto an empty space, as in regular checkers. A jump consists of jumping a checker over an opponent's checker that is in a square one space diagonally in front of the checker to an empty space directly on the opposite side of the jumped checker, as in regular checkers. When an opponent's checker is jumped, it is removed from the board. On a single turn, one checker can jump several of an opponent's checkers in sequence if each jumped checker has a vacant space directly beyond it.

3. A checker is "crowned" or turned into a "King" when it reaches the furthest row of the game board. Two checkers are placed on top of each other to make a King. A King can move and jump forward or backward.

4. If on a turn a player can jump an opponent's checker, the jump must be taken.

5. **Scoring:** A player acquires points when he or she moves a checker from one square to another. The number of points acquired equals the number on the square onto which the checker is moved. If a player jumps an opponent's checker during a turn, the player is awarded points equal to the number on the square onto which the checker lands after jumping. If a player makes repeated jumps with a single checker during a single turn, the player is awarded points equal to the sum of the numbers on the squares on which the jumping checker landed after each jump. Players keep track of their cumulative score throughout the game by adding new points obtained during a turn to points obtained during previous turns. This method of scoring requires a playing strategy that is different from regular checkers.

6. The game ends either after all of one player's checkers have been jumped and removed from the board or after some preset time period (for example, five minutes).

▥ Winning

The player with the highest cumulative score at the end of the game wins.

▥ Playing Variation

➤ Have players write down all of their scores and cumulative scores, in sequence, so that their addition can be checked.

▥ Skill Variation

➤ **Addition, Subtraction, Multiplication, Division:** Any type of problem can be put in the squares. The value of each square is the answer to the equation on it. One- or two-digit whole number addition, subtraction, multiplication, or division problems can be used, as well as problems with fractions, decimals, or integers.

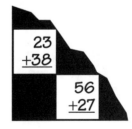

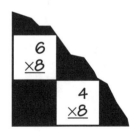

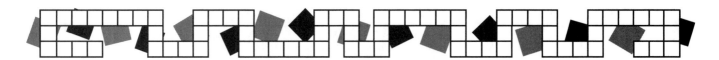

Down the Tubes

SKILL AREAS:
place value, addition, subtraction

Object: Players move their playing pieces along the squares of a game board based on a hundred's number chart, according to the throw of a die, to see who will reach the Finish box first.

Number of Players: two, three, or four

Materials: one die, 4 different colored playing pieces that will fit in the number squares on the game board, 16 index cards or small rectangular pieces of cardstock, and a large piece of white paperboard

Preparation: Copy the game board onto the large piece of paperboard. (The game board can be enlarged on a photocopying machine and pasted onto the paperboard.) Circle in red all numbers with a five in the ones place. Decorate the board, if desired. Prepare sixteen cards, two cards with each of the following numbers on one side: +5, –5, +10, –10, +9, –9, +11, –11.

+5

⊞ Playing

1. To start, each player chooses a playing piece and places it on the 0 space on the game board. The cards are shuffled and placed face down in a pile.

2. Players take turns throwing the die and moving their playing pieces along the game board the number of spaces designated by the die, starting from where they were located at the end of the previous turn.

3. When players move their pieces onto the dotted space at the end of each row, they pretend to slide the piece down the tube to the first space in the next row. The move is not interrupted as a result of slipping down the tubes to the next row of playing spaces. For example, if a player is on space 8 and

throws a 4 on the die, the player moves to 9, moves one more space, slides down the tubes to 10, then moves on to 11, and finally places the playing piece on space 12 where it stays.

4. If a player's piece comes to rest at the end of a move on a number with a 5 in the ones place (the numbers circled in red), the player must pick the top card from the pile and follow its instruction. "+5" means move ahead 5 spaces; "–5" means move back five spaces. If, as result of picking a card, a player moves back beyond 0, the player restarts the game from 0 on the next move. If, as a result of picking a card, a player lands on another circled number, the player stays there and does not pick again. After moving, the player puts the card face down on the bottom of the card pile.

⊞ Winning

The first player to reach the Finish box wins.

⊞ Playing Variations

➤ Use two dice instead of one.

➤ Circle additional spaces.

⊞ Skill Variation

➤ **Factors:** On the playing cards, write statements such as "Go forward to the next prime," "Go back to the last composite number," "Go back to the last number that was a multiple of 3," or "Go forward to the next number with both 2 and 3 as factors."

Down the Tubes

| 0 | 1 | 2 | 3 | 4 | 5 | 6 | 7 | 8 | 9 | 10 |

| 10 | 11 | 12 | 13 | 14 | 15 | 16 | 17 | 18 | 19 | 20 |

| 20 | 21 | 22 | 23 | 24 | 25 | 26 | 27 | 28 | 29 | 30 |

| 30 | 31 | 32 | 33 | 34 | 35 | 36 | 37 | 38 | 39 | 40 |

| 40 | 41 | 42 | 43 | 44 | 45 | 46 | 47 | 48 | 49 | 50 |

| 50 | 51 | 52 | 53 | 54 | 55 | 56 | 57 | 58 | 59 | 60 |

| 60 | 61 | 62 | 63 | 64 | 65 | 66 | 67 | 68 | 69 | 70 |

| 70 | 71 | 72 | 73 | 74 | 75 | 76 | 77 | 78 | 79 | 80 |

| 80 | 81 | 82 | 83 | 84 | 85 | 86 | 87 | 88 | 89 | 90 |

| 90 | 91 | 92 | 93 | 94 | 95 | 96 | 97 | 98 | 99 | 100 |

Finish

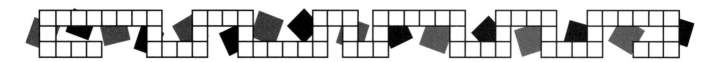

Hit

SKILL AREAS:
multiplication, addition

Object: Players slap two coins so that they slide across a table onto a target. The two numbers on which the coins land are multiplied for a score.

Number of Players: two to four

Materials: two coins or counters, tape, and paper

Preparation: Photocopy the target on the next page or draw a target as shown on a sheet of paper (8.5 by 11 inches). Tape the target onto a table, about 12 inches from an edge. That edge will be the hitting edge. Mark or tape arrows on the coins. (It's important that the paper is taped smoothly, so coins will not get stuck underneath.)

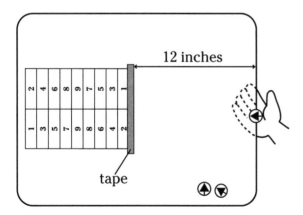

▦ Playing

1. There are 10 innings in the game. Each player takes one turn during each inning.

2. A player places a coin on the hitting edge of the table, as shown. The player then gives the coin a sharp, light slap with the palm of the hand so that the coin slides onto the target. The player repeats this with the other coin.

3. The player multiplies the two numbers on which the arrows of the coins land. For example, if the arrows land on 5 and 9, the player multiplies 5×9 and gets the product 45. This is the player's score for the turn. If an arrow lands off the target, the player's score is zero for the turn.

4. A player completes a turn by adding the new score to the sum of previous turns. (Players should carefully check each other's calculations.)

5. The game ends after 10 innings.

▦ Winning

The player with the largest total score wins.

▦ Playing Variations

➤ Put the target in a box top or on the floor and have players toss the coins onto it.

➤ At the end of each inning compare the products players get on their turns and award one point to the player with the largest product. The player with the most points at the end of the game wins.

➤ Reduce the number of rounds.

➤ Make different types of targets. For example, make a circular target.

▦ Skill Variations

➤ **Addition:** Play as an addition game.

➤ **Fractions, Decimals, Place Value:** Put other numbers on the target, such as fractions, decimals, or three-digit numbers.

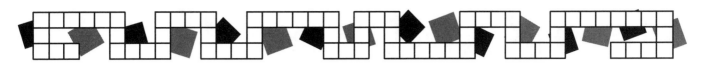

1	2
3	4
5	6
7	8
9	9
8	7
6	5
4	3
2	1

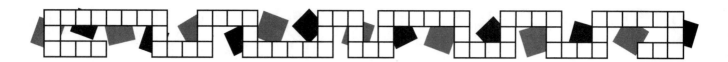

Drop the Die on the Donkey

Object: Blindfolded players toss a die on a game board that is a picture of a donkey, and score points equal to the product of the die number and the value of the part of the board on which the die lands.

Number of Players: two to five

Materials: a square piece of paperboard, one die (or wood cube), blindfold, and colored markers for making and decorating the game board

Preparation: Copy the game board onto the paperboard and decorate it, if desired. To practice facts other than those from 1 to 6, use a wood cube numbered as desired instead of a traditional die.

▦ Playing

1. The game is played as a series of rounds, during which each player takes a turn and scores points. Players obtain cumulative scores by adding together points from successive rounds.

2. During a turn, a player first puts on the blindfold. The die is placed in the player's right hand. Other players in the game rotate the game board on the floor in front of the player. The first finger on the player's left hand is then placed on one corner of the game board, so that the player can locate where it is. The player raises the hand holding the die about five inches above the game board and drops the die. The blindfold is then removed.

3. A player's score for a turn is the product of the number on the die and the number in the playing board area on which the die landed. If the die touches two different areas on the playing board, the player can take the higher

value of the two areas, even if only a small corner of the die is touching the area. The player adds the score from the current turn to the sum of the scores from previous turns to get a cumulative score.

▦ Winning

The player with the highest cumulative score after a set amount of time or number of rounds wins.

▦ Playing Variation

➤ Play as the traditional Pin the Tail on the Donkey, only have a blindfolded player select a numbered tail from a grab bag to start a turn. Score as above.

▦ Skill Variations

➤ **Addition 1:** Have players add the number on the die to the number on the playing board. Play each round as a separate game.

➤ **Addition 2:** Use different numbers on the playing board. Numbers less than 20 are good for a game in which the first player with a cumulative sum of more than 80 wins. Numbers less than 200 are good for a game in which the first player with a cumulative sum of more than 800 wins. Use a button instead of a die, and a player's score for a round is equal to the number in the area of the game board on which the button lands. Cumulative scores are still determined by adding together scores from successive rounds.

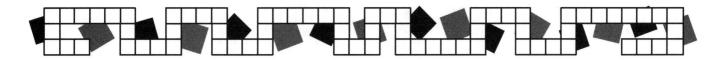

Drop the Die on the Donkey

Divi

Object: A dealer takes a handful of counters (n) and throws a die (d). Players bet on the integer result of dividing the number of counters by the die number (n÷d) and the number of counters remaining after the division.

Number of Players: three or more

Materials: about 60 small counters such as buttons, pebbles, or dried beans; a container to store them in; a wood cube or die; paperboard on which a game board can be constructed; paper, and pencils

Preparation: Number the sides of the cube from one to six, if it is not a conventional die. Make an enlarged copy of the game board (at least 12 inches per side); laminate and decorate it as desired.

▦ Playing

1. Divi is played in rounds. To begin each round the dealer takes a handful of counters and places them in a pile in the center of the game board, rolls the die, announces the number thrown on the die, and asks players to place their bets. Two bets are placed. One bet is for the number of groups that will be created when the counters are divided into groups, each of which have as many counters in it as the number thrown on the die. The other bet is for the number of counters that will remain at the end of the division process. Players record their bets on a sheet of paper.

2. After all bets are placed and recorded, the dealer separates the counters into groups, each of which has a size equal to the number thrown on the die. Each group of counters is placed in one of the small squares on the game board. When no more groups can be

formed, the dealer announces the number of groups created and the number of counters remaining.

3. Players score one point for guessing the number of groups correctly and one point for guessing the remainder correctly.

▦ Winning

The player who has the largest number of points after a specified number of rounds or a specified time period wins.

▦ Playing Variations

➤ Bet just on the remainder or bet just on the number of groups.

➤ Do not use the die and have a predetermined size of the groups for the entire game. The number 4 works best and the numbers 3 and 5 work well, too.

➤ If playing with a large group use an overhead projector as the playing surface and project the division process.

➤ Have older players record the division problem once they know the number of groups, die number, and remainder.

▦ Skill Variation

➤ **Place Value:** Guess the number of counters put on the table. A correct guess gets two points, if either the ones or tens digit is correct, the player gets one point.

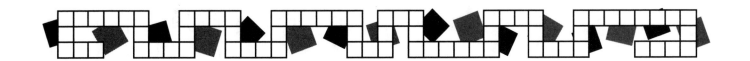

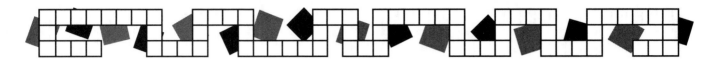

Race

Object: Players select slips of paper from a bag. Each slip contains an arithmetic problem and a distance to move on the game board. Players more accordingly.

Number of Players: two to four

Materials: paperboard, 4 small objects for playing pieces; a paper or cloth bag about 8 inches wide by 12 inches high, and 50 to 100 small index cards or printer cards

Preparation: Copy the game board on the next page onto the paperboard. Decorate as desired. Laminate or cover the board with clear contact paper. On each of the cards write an equation and directions to move 1 to 5 spaces as shown. Match moves of 1 or 2 spaces to easy problems and moves of 3, 4, or 5 spaces to more difficult problems. The problems put on the cards can correspond to any skill or operation. Prepare between 50 and 100 cards for the game.

$4 \times 5 =$

move 2 spaces

Playing

1. Each player chooses a playing piece and places it on the start position. The cards with arithmetic problems are placed in the bag, the bag is closed, and the cards are mixed up by shaking the bag. Players determine the order of play.

2. The first player randomly picks a card from the bag and attempts to solve the problem on the card. If successful, the player moves his or her playing piece toward the finish line the number of spaces designated on the card. If the player does not answer the problem correctly, the piece stays put. At the end of each turn, the player returns the card to the bag and shakes it up.

Winning

The first player to reach the finish line wins.

Playing Variations

➤ Place penalty cards in the bag that require the player to move backward a certain number of spaces.

➤ The spaces on the playing board can be numbered so the players can more easily remember where their pieces are.

➤ Ask players to record on paper the arithmetic problems they draw from the bag and their answers.

Skill Variations

➤ **Money:** Stamp coins on a card and have players calculate how much money is represented on the card.

➤ **Fractions:** Draw fractional amounts on the cards, and have players state the amount represented.

➤ **Factors:** Place one or several numbers on a card, and have players state the factors of the number (or the common factors of the several numbers).

Race

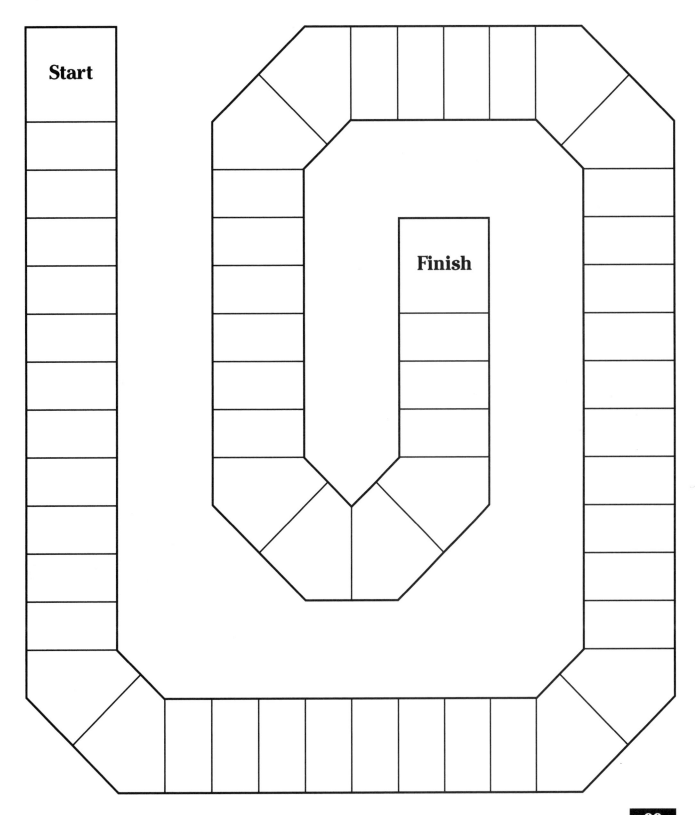

Start

Finish

Physical Education Games

Physical education games have long been a favorite of children. Kids love any chance to run, jump, and play. And for those children who learn best through large motor activity, physical education games are an ideal way to help them learn mathematics.

Most physical education games need a large empty space, such as a playground or gym. Many can be played in a hall. Some of the best can even be played in a classroom, as long as there is sufficient room. Always make sure that the playing areas in which physical education games take place are safe for children and that an accidental fall will not result in a child colliding with a desk, door, or wall.

Jump the Answer

SKILL AREAS: addition, subtraction, multiplication, division

Object: Players race to be the first to jump answers to problems on a number grid.

Number of Players: two teams of any size

Materials: chalk or crayons, chalkboard or sheet of paper upon which problems can be written, paper, and pencils

Preparation: Draw number grids and starting lines (as shown) on a floor, sidewalk, or playground. Squares in the number grid should be about one foot long. The distance between the starting line and the number grid should be between eight and twenty feet.

7	8	9
4	5	6
1	2	3
	0	

7	8	9
4	5	6
1	2	3
	0	

_____

starting line

✋ Playing

1. Choose a problem giver and two teams. The players of each team line up behind their respective starting lines.

2. The problem giver writes a problem on the chalkboard or a sheet of paper and shows it to the first players on line. The problem can involve any operation with any size numbers, so long as it has a whole number solution (for example, addition problems can vary from $7 + 5 = ?$ to $3567 + 8456 = ?$).

3. The two players calculate the answer (either by using mental math or with paper and pencil) and then run to their number grid and jump the answer on the grid, with digits being jumped according to their position in the number from highest place value position to lowest place value position (for example, if the answer is 36, the student would first jump to the 3 and then to the 6). Students can jump into the squares with either both feet or on one foot. Once the answer to the problem is jumped, the player races back to the starting line.

4. The player who returns to the starting line first, wins one point for his or her team. Players waiting in line for their turn must calculate answers to the problems and watch to make sure that the other team's players jump the correct answers.

5. The players who just jumped go to the back of the line after the point is awarded and the next players move up to the starting line to get ready for the next problem. The process repeats until each player has a turn to jump.

✋ Winning

The team with the most points at the end of the game wins. The game can end after a specified number of points has been acquired or a set time period.

✋ Playing Variations

➤ Simple problems can be spoken aloud rather than written.

➤ Players can be given numbers instead of equations to jump.

Math Ball

Object: Players stand in a circle and toss a ball to each other. As a player catches the ball, he or she must say the number one higher than that said by the previous person who caught the ball.

Number of Players: three or more

Materials: a ball, about four to 12 inches in diameter

Preparation: No preparation needed

Playing

1. Players stand in a circle. The size of the circle depends upon the skill level and number of players.

2. The first player, who is the one holding the ball at the start of the game, yells "zero."

3. Then he or she tosses the ball to any other player in the circle (the catcher). Before catching the ball, the catcher must yell the number one higher than that yelled by the tosser.

4. If the catcher does not yell the appropriate number, does not yell the number before catching the ball, or does not catch the ball, that player must drop out of the game and stand outside of the circle. The ball is then passed back to the tosser who throws it to another player.

5. If the tosser does not throw the ball to the catcher in a way that the catcher can catch it, the tosser must drop out of the game and the designated catcher restarts the game's count where the tosser left off.

Winning

The last player left in the game wins.

Playing Variations

➤ Play as a counting down game with the first number being 100 (or 1000).

➤ Start the count at a number other than zero.

➤ Slow the game down by ruling that the ball must bounce on the ground once between the tosser and the catcher.

Skill Variations

➤ **Multiples:** Count by a multiple other than one.

➤ **Fractions:** Count by adding a fraction to the previous number yelled.

➤ **Buzz:** Play like Buzz, with the ball being passed clockwise, and with players having to say "buzz" (instead of the number) whenever the multiple of some designated number is to be said. The word "buzz" can also be used as a substitute for any digit in a number equal to the multiple (thus for multiples of 5, the number 55 would be said as "buzz-buzz").

"Simon Says" Math

SKILL AREAS: addition, subtraction, multiplication, or division

Object: Similar to Simon Says. Simon asks players to perform tasks involving mathematics, such as jump the answer to $42 \div 6$.

Number of Players: three or more

Materials: none

Preparation: No preparation needed.

Playing

1. Choose a leader. Have all players stand in an area where they can move about freely.

2. The game is played as a series of rounds. The leader begins each round by making a statement that has three parts: a Simon Says part, an action part, and an equation part. The Simon Says part either contains the words Simon Says, or nothing at all. The action part describes some physical action that players are to engage in. The equation part of the statement gives an addition, subtraction, multiplication, or division equation. Different operations can be used in a game. Two examples of statements are "Simon says jump the sum of 3 plus 2." and "Clap your hand to show the number of tens in the product of 7 times 6."

3. Once the leader has made the three-part statement, players carry out the action described in the statement if the statement begins with the words "Simon says." If the statement does not begin with these words, the players stand still.

4. A player drops out of the game and sits down if: the player stands still when a statement beginning with "Simon says" is made; the player acts out all or part of the answer to the equation when the statement lacks the phrase "Simon says"; the player performs the described action incorrectly (for

example, hops on the right foot when directed to use the left); or if the player incorrectly acts out the equation. Such errors are monitored by the leader and other players.

5. A player stays standing and remains in the game if the action was performed correctly.

6. A round ends when it is determined which players will stay in the game and which must drop out. A new round then begins.

Winning

The last player or players to remain standing after some specified period of time win.

Playing Variations

➤ Write actions on one set of 3×5 index cards and equations on another set of cards. Have the leader pick one card from each set to determine what is to be said. Here the leader simply decides whether or not to add "Simon says."

➤ Entire statements can be written on index cards. The leader then just has to shuffle the cards to begin the game, pick a card, read it, and help determine who remains in the game and who drops out.

Skill Variation

➤ **Place Value:** In place of an equation, the leader says a two-, three-, four-, five-, or six-digit number. In the action part of the statement the leader specifies if the players are to act out the digit in the ones, tens, hundreds, thousands, etc. place in the number. For example, the leader might say, "Simon says clap your hands the number of tens in 475."

Number Calisthenics

Object: Two teams take turns doing a series of exercises to instructions related to place value. The place value of the digits of a number determines how many of which exercises will be done.

Number of Players: two teams of any size

Materials: none

Preparation: No preparation needed.

Playing

1. Choose two teams and a leader to announce numbers. The two teams alternate turns as exercisers and watchers.

2. A turn begins with the leader announcing (and/or writing on a board that all players can see) a four-digit number. The exercisers must then act out the four-digit number in unison by doing in sequence: as many jumping-jacks as specified by the value of the digit in the thousands column of the number; as many touch-your-toes as specified by the value of the digit in the hundreds column; as many clap-your-hands as specified by the tens digit; and as many snap-your-fingers as specified by the value of the digit in the ones column. For example, for the number 5,432 the exercisers would do 5 jumping-jacks, 4 touch-your-toes, 3 clap-your-hands, and 2 snap-your-fingers.

3. While the exercisers are acting out their number, the watchers watch them to see if anyone makes a mistake. Anyone who makes a mistake exercising must drop out of the game. Players who are out of the game become permanent watchers for their team.

4. Once one team has completed exercising to a number, the teams switch roles and a new number is announced by the leader.

5. Play continues until either all the members of one team are out, until the amount of time allocated for playing the game elapses, or until a specified number of rounds have been played.

Winning

The team that has the most players still in the game when it ends, wins.

Playing Variations

➤ Have the leader write the number rather than saying it aloud.

➤ Replace any of the exercises with others. For example, the slap-your-hands might be replaced by squat-thrusts.

➤ Any sized number can be used: two-, three-, four-, or five-digit numbers all work well.

➤ This can be played as a whole-group activity where there are no winners and no losers.

Fingers

Object: Two players face each other with one fist closed. Simultaneously, they extend from one to five fingers while also calling out a number between 2 and 10. If a player's called number equals the sum of the extended fingers, that player wins a point.

Number of Players: two

Materials: none

Preparation: No preparation needed.

Playing

1. Fingers is played as a series of rounds.

2. During each round, two players face each other with one of their fists closed. On the count of three, they both simultaneously extend from one to five fingers while at the same time calling out a number between one and ten. Fingers are extended with a shake of the fist, in a manner similar to that used in the game Paper-Scissors-Rock.

3. If the number called out by a player equals the sum of the fingers extended by both players, that player scores one point. Both players can score a point during the same round.

4. A player's score is the sum of the points acquired during the game.

Winning

The player with the highest score wins.

Playing Variations

➤ Allow fewer than five fingers to be extended. This allows the probability of guessing correctly to be higher.

➤ Allow players to use two hands when extending fingers, so that numbers up to 10 can be used.

Skill Variations

➤ **Subtraction:** Have players bet on the difference between the numbers of fingers extended, by calling a number between 0 and 4.

➤ **Multiplication:** Have players bet on the product of the number of fingers extended. Bets that are within four numbers above or below the product win a point (thus a bet of 15 will win a point if the product is 12, 15, 16, or 18).

Hands In

Object: Each player hides from one to three small counters in their left hand and guesses the sum of the counters hidden in all players' hands.

Number of Players: two to ten

Materials: three small counters such as pennies, paper clips, dried beans, buttons, or pebbles for each player

Preparation: No preparation needed.

Playing

1. Each player is given three small counters at the start of the game. The game is played as a series of rounds.

2. Players start each round by hiding both of their hands and their counters (in their lap, under a table, behind their back) from other players; placing one, two, or three counters in their left hands; and concealing the remaining counters from the other players in their right hands. On the count of three, all players extend their left hands.

3. Players guess, one at a time in clockwise rotation, the total number of counters contained in the outstretched hands. Once all players have guessed, they open their hands and calculate the total number of counters (by either adding or counting them).

4. At the end of each round players receive points. Players who guess correctly score two points. If no player guesses correctly, the player (or players) with the closest guess score one point.

5. A player's cumulative score is the sum of the points received during the game.

Winning

The player with the highest cumulative score wins.

Playing Variations

➤ Change the number of counters given to each player. Two to five counters can be used.

➤ Have the players record their guesses on a sheet of paper during each round so that there is no doubt about what they guessed.

Twist-'em

Object: Players give each other a four-digit number which they must represent by placing their hands and feet on a numbered grid. The player who falls over first while attempting to represent the number loses.

Number of Players: two to four (or two larger teams)

Materials: chalk or crayons for drawing on the floor or ground, paper, and pencils

Preparation: Draw the numbered grid as shown on a floor, sidewalk, or playground. The side of each cell should be from eight to fifteen inches long, depending upon the size and skill of the players.

7	8	9
4	5	6
1	2	3
	0	

✋ Playing

1. Players take turns as number giver and twister.

2. The number giver writes a four-digit number on a sheet of paper and reads it aloud to the twister using correct place value language. The twister checks the paper to make sure the number giver read it correctly.

3. The twister then puts his or her hands and feet in the appropriate cells of the grid in the following order:

   ➤ right foot on the digit representing the thousands,

   ➤ left foot on the digit representing the hundreds,

   ➤ right hand on the digit representing the tens, and

   ➤ left hand on the digit representing the ones.

4. After getting into position, the twister tries to count from 1 to 10 without falling.

5. If the twister falls while attempting to get into position and count from 1 to 10, the players switch roles. If the twister succeeds in the endeavor, players switch roles.

6. Score as follows: Number giver gets one point for reading the number correctly. Twister gets five points for counting to 10 and not falling. Players must have an equal number of turns in the roles of number giver and twister.

✋ Winning

The player with the highest score wins. Equal scores produce a tie.

✋ Playing Variation

➤ If playing with two teams, the game moves faster and is more exciting if two number grids are used and each team simultaneously gives the other team a number and receives a number to act out on their own grid.

✋ Skill Variation

➤ **Addition or Subtraction:** Addition or subtraction problems involving four-digit numbers can be used, with the twister solving the problem before acting out the answer. A correct solution to the problem earns the twister two points.

Math Rover

Object: A version of Red Rover. Players with numbers attached to them run across a playing field while a person who is "It" tries to tag them.

Number of Players: 7 to about 30

Materials: a placard for each player (piece of paper or cardboard) about 8 by 11 inches, string, and a large playing field

Preparation: On each of the placards write a different number from 1 to x, where x is the number of signs being created. Make holes in each sign and attach string in such a way that it can be hung around a player's neck or tied around the player's waist. Mark out a rectangular playing field that is about 20 feet by 40 feet. A blacktop is ideal.

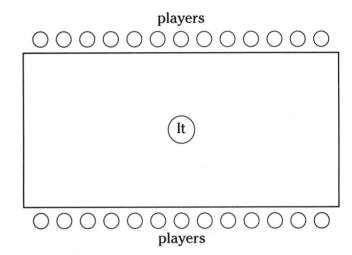

players

It

players

🖐 Playing

1. To begin, each player puts on a number placard. One player is chosen to be It. It stands inside the playing field. The rest of the players are divided into equal groups and stand along the outside of the longest sides of the field.

2. It decides which player is going to run from one side of the playing field to the other and constructs an addition or subtraction problem that is equal to the number on that player's placard. It then sings out the chant "Math Rover, Math Rover, let (state the problem) come over!" (for example, "Math Rover, Math Rover, let 9 minus 4 come over!" forces 5 to run). After the chant, the player with the answer to the problem attempts to run from one side of the rectangle to the opposite side without being tagged. It attempts to tag the player while he is on the inside of the playing field.

3. If the runner reaches the other side without being tagged, that player is safe and waits for his or her number to be called again. If the runner is tagged or does not run because he did not calculate correctly, that player goes to the center of the rectangle and becomes It's assistant, helping tag other players. A player who runs across the field when it is not his or her turn must enter the rectangle and become an assistant to It.

4. The game continues with It calling and players running (repeating steps three and four) until only one person is left who has not yet been tagged.

Winning

The winner is the last untagged player.

Playing Variations

➤ An adult can construct and call problems out so that It and his or her assistants need only tag players.

➤ Problems can be put on index cards, with two or three problems for each placard. It picks cards one by one and uses their problems in the Math Rover chant. If a tagged player's number is called, another card is picked.

➤ Use only the numbers from 1 to 10 on the placards, so that several players have the same number. When It calls a problem, all players with the answer must run. One or more might be tagged. Only those tagged become assistants to It.

Skill Variations

➤ **Inequalities:** It can call out inequalities that force more than one player to run. (Examples: "All players greater than $5 - 3$ and less than 8×2" or "All players less than $8 + 2$")

➤ **Division:** It calls out division problems. Placards contain only the numbers from 1 to 10, and more than one player might have to run when a problem is called. Players run if either the quotient or remainder equals their number.

➤ **Equivalent Fractions:** Put fractions on the placards. It calls out fractions that are equivalent to, but different from, those on the placards.

Operation Hopscotch

Object: Similar to hopscotch, but before each hop, players call out the product of two numbers.

Number of Players: two to four

Materials: chalk or crayons, and a small stone for each player

Preparation: Draw the numbered grid as shown on the floor, sidewalk, or playground.

```
      10
    8    9
      7
    6    5
      4
    2    3
      1
      0
```

Playing

1. Players take turns. To start the first turn, a player stands in 0 and throws his or her stone into square 1. He or she jumps over 1 into 2 and 3 (with one foot in 2 and the other in 3), hops onto 4 (with one foot), jumps into 5 and 6, hops onto 7, jumps into 8 and 9, hops into 10, turns around in square 10, and goes back the same way, jumping over square 1 and back into square 0.

2. Each time a player puts a foot on a number or numbers, and before hopping or jumping to succeeding numbers, the player must call out the product of that number (or numbers) and the number on which his or her stone presently resides (for example, a player whose stone resides on 2 would call out the following while jumping, "2, 6, 8, 10, 12, 14, 16, 18, 20, 20, 18, 16, 14, 12, 10, 8, 6, 2, 0"). If a product is incorrect, the player loses that turn and must start over and toss the stone into the same numbered square on his or her next turn.

3. After a player correctly gets to back to 0 or loses a turn, the next player takes a turn.

4. If during a turn a player touches any of the lines with a foot, puts a foot down when hopping, loses balance and falls, or throws his or her stone out of the desired square, that player loses that turn and must start over on the next turn from where they were previously.

5. Players must always jump or hop over any square containing their stone or an opponent's stone.

6. On the next turn, once a player succeeds in completing a turn with their stone in square 1, a player picks up the stone from square 1, throws it into square 2, and repeats the hop and jump sequence described above. On future turns this process is repeated, the stone being thrown into the next square in the sequence.

Winning

The first player to complete a turn with their stone in square 10 wins.

Playing Variation

➤ Have players pick up their stone just before they jump over it on their return from 10 to 0. This way there is only one stone on the playing board at a time.

Skill Variation

➤ **Addition:** Players add the number on which they are going to jump to the number on which their stone presently resides.

Bean Bag Toss

SKILL AREAS:
multiplication, use of a
multiplication table, addition

Object: Players take turns tossing a bean bag at a target containing a blank multiplication table. Players score points equal to the answer to the multiplication problem they toss.

Number of Players: two to five

Materials: a bean bag, a large piece of cardboard, a marker, and a piece of tape

Preparation: Use the marker to draw a multiplication table on the cardboard, leaving out the answers. This will be the target for the game. A target for the numbers 1 to 5 is shown. Targets for the numbers from 1 to 10 can be constructed. Place the target on the floor and put a two-foot strip of tape on the floor about 10 feet from the target to indicate a throw line.

×	1	2	3	4	5
1					
2					
3					
4					
5					

Playing

1. During a turn, a player stands behind the throw line and tosses the bean bag toward the target. If the bean bag does not land on an empty cell in the target, the player gets a second chance to throw the bean bag. If the bean bag still does not land on an cell of the target, the player scores zero points for that turn.

2. If the bean bag lands on an empty cell in the target, the player multiplies the number in the horizontal row by the number in the vertical column. The answer to that problem is the player's score for that turn, if the player does the multiplication correctly. If the player does the multiplication incorrectly, the player scores zero points for that turn. If the bean bag lands so that it is on several blank squares simultaneously, the player can choose which problem to solve.

3. Players keep a record of their scores by adding their scores from each round.

Winning

The player with the highest cumulative score after each player has had the same number of turns wins.

Playing Variations

➤ The target can be drawn in a large cardboard box and players can toss a paper clip instead of a bean bag.

➤ Players can be allowed to toss the bean bag until it lands on a blank cell on the target.

➤ Scores can be calculated on a round-by-round basis with the player with the largest product receiving only one point for the round.

➤ Allow players to use a calculator to check questionable answers.

Skill Variations

➤ **Addition:** Use an addition table rather than a multiplication table.

Randomizing Devices

Number Cubes   Cover the faces of regular dice with masking tape and writing on the new symbols, or by marking blank cubes purchased from an educational materials supply company. The size recommended is between ¾ and 1 inches (or 2 and 2.5 centimeters).

Alternately, use the pattern at right to create your own cubes. Number the six faces. Then cut along the solid lines and fold on the dashed lines. Glue or tape the flaps behind the faces to secure.

Pencil   Mark the six sides of a pencil, roll it, and see which side faces up when the pencil comes to rest.

Teetotum   Cut a hexagon about 3 inches (7 cm) across from stiff cardboard and divide it into six equal triangles. Write the desired mathematical symbols in the triangles and stick a pencil point down through the center. Spin the teetotum; when it stops the symbol in the triangle resting against the playing surface is "it."

Spinner   Draw a circle on cardboard and dividing it into six wedge-shaped sections. Place a piece of tape over the center and attach a short pencil with a thumbtack pushed

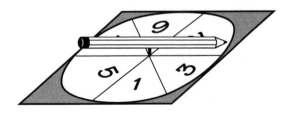

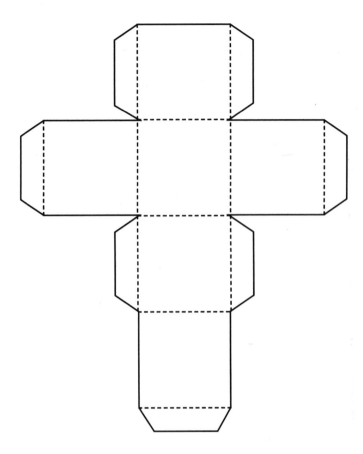

through the circle's center from the back. Try to impale the pencil approximately in the middle. Spin the pencil to get the number or symbol for play.

Tossing Board   Divide a sheet of paper into six sections and write a symbol or number in each. Toss a button or pebble from a short distance away and use the symbol it lands on for play.

4
6
2
5
1
3